Message from Global Leaders

Abhik Chaudhuri's book reveals the enormous new power of connecting millions of people, sensors and intelligent machines into what my colleague Thomas Malone calls "Superminds." This is the age of the IoT. Abhik Chaurdhuri explains how these new IoT-enabled minds are transforming our homes, cities and society faster than most of us realize.

The book accomplishes the near-impossible, offering a helicopter's view of the new landscape: It provides rich ground-level descriptions of key IoT technologies and systems. We grasp that these new enablers are real, cheap and plentiful. The book interweaves description with systematic overviews, offering wise counsel on how this profusion can be brought together to solve our most critical problems. A smart business executive or policy maker will benefit immensely from reading this book. Opportunities abound, and Chaudhuri enables us to consider them well.

James F. Moore
Strategist
Collective Intelligence Design Lab
Center for Collective Intelligence, MIT
Inventor of the business ecosystems approach to strategy-making

The Internet of Things (IoT) is one of the most important and complex technologies of our 21st-century digital economy, which promises to make our homes, cities, factories, vehicles, health systems and the world around us smarter and more responsive. The IoT is not an individual technology, but rather an ecosystem of products, services and private and public sector institutions, all of which must work well together to achieve its potential. But, like any advanced technology, the IoT must be properly managed to realize this potential while ameliorating its accompanying security and privacy risks as well as serious unintended consequences.

Abhik Chaudhari's Internet of Things, for Things and by Things is a timely contribution to help achieve the promise of the IoT. His comprehensive book explains the fundamentals of the IoT, its potential applications, inherent risks and governance requirements to anyone interested in better understanding the IoT and its implications for economies and societies.

Irving Wladawsky-Berger
Emeritus, IBM; Visiting Faculty, MIT

Abhik Chaudhuri has researched and penned the most comprehensive compendium of information and insight on the IoT that I am aware of. He leaves no stone unturned—from theory and philosophy of information in the context of the IoT, to its ever-present security and privacy concerns and dozens of practical use-cases and real-world examples. Anyone looking to learn almost all there is to learn about the IoT, or about specific aspects of the IoT, should strongly consider this book.

Douglas Laney
VP and Distinguished Analyst, Gartner,
and author of the book Infonomics: How to Monetize, Manage,
and Measure Information for Competitive Advantage

Few thought leaders have Abhik Chaudhuri's spectacular analytical ability to convert the complexity of the Internet of Things into an easy-to-comprehend book that expands and more deeply explains the original concept behind the IoT, which I devised in 1985. As this technology exponentially grows, it is important that publications of this nature are widely distributed so that other forward thinkers, like Abhik, can contribute to the more advanced integration of IoT technology with the decentralized features of blockchain, device data transportation improvements via 5G wireless and the emergence of complex systems, like unmanned aircraft systems traffic management, which will have a profound impact on the conduct of more efficient operations in our lives.

Peter T. Lewis
President, NetMoby, Washington DC
Inventor of the term 'Internet of Things' and the concept behind it

Privacy by Design (PbD) is a framework I created in the late 1990s for preventing privacy harms by proactively embedding the necessary privacy-protective measures into the design of information technology, networked infrastructure and business practices. PbD is a model of prevention—prevent the privacy harms from arising. The Internet of Things, or perhaps more aptly, the Internet of Everything truly concerns me because the interconnected nature of virtually all that we do may lead us down a path of surveillance that will be too great to conquer after the fact. Surveillance is the antithesis of privacy, and accordingly, the antithesis of freedom. I strongly believe that neither privacy nor the benefits inherent in the emerging IoT need to be sacrificed. If we are to preserve any semblance of privacy in such an emerging world, we must ensure that privacy is built into the very systems being developed. This will require innovation and ingenuity, as well as foresight and leadership, in an effort to reject unnecessary trade-offs and the false dichotomies that jeopardize privacy in favor of other laudable objectives. We can have privacy AND the Internet of Things—but only if we act now and proactively embed Privacy, by Design!

Ann Cavoukian, Ph.D., M.S.M.
Distinguished Expert-in-Residence
Privacy by Design Centre of Excellence
Ryerson University

This comprehensive book that Abhik Chaudhuri has authored will be an important contribution to many parties. He covers perspectives from the technical to the socio-technical and to important areas such as smart cities, governance and standardization. The book provides a synthesis from many sources including Abhik's own creative contributions.

Given the complexity of the multitude of technical and non-technical issues involved, this book should be a starting point for those who seek to work towards understanding and addressing the complexities and thus seeking solutions to the many opportunities and risks involved in the new world environment created via the Internet of Things.

Certainly deploying a systems perspective in considering the many interacting elements of the system of systems involved will be vital.

Harold "Bud" Lawson, Professor Emeritus
Coordinating Editor of the College Publications Systems Series
ACM, IEEE and INCOSE Fellow
IEEE Charles Babbage Computer Pioneer
INCOSE Systems Engineering Pioneer

The Internet of Things, for Things and by Things by Abhik Chaudhuri successfully illustrates all the different aspects related to the IoT, such as technology, people, philosophy and regulations. The book captures everything related to the IoT, from architecture, the philosophy of information, ethics in the IoT and privacy issues. Moreover, the book touches the aspects of threat mitigation and risk management of interdepended systems of smart cities, smart city governance and regulations. This book manages not only to provide theoretical information about the IoT but also provides examples of IoT technology and potential benefits to society in various domains. The reader can sufficiently understand the complexities of the IoT ecosystem. I recommend this book as a valuable addition in the field of the IoT not only for experts but also for everyone who will be using IoT technology in the future.

Dr. Maria Bada
Senior Researcher at Global Cyber Security Capacity Centre University of Oxford

Abhik Chaudhuri has assimilated and asserted the triple-foundations of the IoT. It is like reaching the foundations of the very democracy of information building, which is OF the architecture and the embedded philosophy, then FOR the array of designs that makes our built environment livable and inclusive! And finally, BY the players and stakeholders of all classes and categories that configure our humankind and that make any application complete. The new and deep course of the information building of our future, which is the IoT, is very well researched by Chaudhuri and it has been done in a unique way, never before and never attempted in the way that one may find while paging through the book. The book will go a long way and find a place on many shelves and in many minds of our better tomorrow.

Dr. Joy Sen
Professor and Head, Department of Architecture and Regional Planning, Indian Institute of Technology, Kharagpur, and author of the book 'Sustainable Urban Planning' (The Energy Research institute)

I see the Internet of Things (IoT) being vitally important for three reasons:

- Firstly the number of devices means that the scale of the IoT will be more pervasive than the 'Internet of People' revolution we have seen to date.
- Secondly, the IoT has many things in common with the way we manage our money digitally. It comes with similar risks and needs good policy and governance to work at all. This means we need to apply a lot of wisdom and integrity in our use of the IoT. There is a lot of opportunity for failure, and success will involve working with care in a complex, rapidly emerging technical environment.
- Thirdly, the data that is produced by the IoT is proving to be a new natural resource on the planet. It gives us the ability to understand the way the world works in finer detail, at a bigger scale and in a more timely manner, thereby increasing our chances of success and avoiding dumb decisions.

If handled well, the IoT will give us the understanding and the foresight to make the world a better place economically, socially and environmentally.

Along with the growth in traditional computing power and emerging cognitive computing, the application of the IoT gives mankind the tools to seriously make progress on our Millennium development goals.

Ian Abbott-Donnelly
Associate: Cambridge University Centre for Science & Policy
Emeritus member: IBM Academy of Technology

The Internet of Things, for Things and by Things provides a comprehensive view of the many opportunities for new value in our hyper-connected communities, from smarter healthcare to connected vehicles. Abhik Chaudhuri wisely includes consideration regarding how to ensure trust, identity, privacy, protection, safety and security in all these systems, from the sensor to the cloud and everywhere in between, to ensure safe and responsible implementation of these new connected systems of systems in our world.

Florence Hudson
"Blockchain in Healthcare Today" Editorial Board, Former IBM Vice President,
Former Internet2 Senior Vice President and Chief Innovation Officer

The world of automotive transport is governed by diverse laws, regulations and standards, and we have no difficulty understanding the purpose and nature of these controls, regardless of how much we like them. Broadly described, the controls apply to the design and manufacture of vehicles across numerous classes, to the design and construction of roads on which vehicles operate and to the use and operation of vehicles by individuals and organizations. In 2003, we realized that the world of information technology needed to be governed by more than the technical standards for computer software, hardware, stored and transmitted data and operational management. From this realization was born ISO 38500, the standard for governance of (the use of) information technology. While stated with excessive subtlety, ISO 38500 positions business leadership and ownership of the use of information technology as a critical element in extracting true value from the investment. The challenge has been for business leaders to understand the role they play in defining the use of information technology and in enabling it to fulfil its promise. Yet these elements are comprehensively captured in the six principles of ISO 38500: responsibility, strategy, acquisition, performance, conformance and human behavior. These principles apply in all cases where information technology is deployed.

The Internet of Things is, at its most primitive definition, an organized, large scale deployment of information technology. Despite extensive current deployments, the IoT remains in its early infancy. As scale and technology evolution make devices and communications inexorably cheaper, the potential extent to which humanity might deploy future Internet of Things structures becomes virtually boundless. Those who conceive and implement such networks will benefit from comprehensive guidance on both the capabilities and the controls that interwork to make an IoT solution both effective and safe.

In this expansive work, Abhik Chaudhuri lays out for all to understand the nature of the Internet of Things, enabling business leaders and technologists alike to explore possibilities, to plan previously inconceivable possibilities and to realize extraordinary value from the opportunity to capture data and exert control in ways far beyond the naked capability of human beings. Before you take another step into your (organization's) plans for the IoT, and even if you think the IoT is not relevant to you, read this book!

Mark J Toomey
Executive Chair, Digital Leadership Institute Limited
ISO Project Editor: ISO/IEC 38500:2008
Author: Waltzing with the Elephant

As a former CIA officer who used to focus on terrorists and now applies those lessons to technology, I'm not a big fan of letting the world into your homes. This is what the IoT does.

But, do I have smart bulbs that automatically turn on when I get home? How could I not? Does Alexa play my favorite songs whenever I ask? Of course.

IoT devices inevitably will be hacked, and eventually on a large scale that will get to you and me. No, your smart coffee maker can't be hijacked to shoot scalding coffee on you . . . yet! But it certainly might listen to your conversations. Or your smart lights might be overheated to explode and shoot glass at you.

Your smart coffee pot isn't going to have antivirus software. Nor will two-factor authentication make sense. Could you imagine having to get a text message and punching in a verification code for your coffee maker? And then for your Alexa? And then for your smart lightbulbs? It's not feasible. The IoT is going to require new cybersecurity controls customized for its unique nature.

I am a big proponent of the IoT. The benefits are enormous. But I urge everyone to keep security in mind. The IoT is like letting the terrorists into your backyard. You don't have to spend time at the CIA to know how that will play out.

Remind IoT manufacturers that cybersecurity is important to you. Push them to be transparent about their cybersecurity practices, and to communicate the level of their cyber strength. And when deciding who to buy from, consider security along with things like features and price.

Scott M. Schlimmer, CISSP
Co-Founder & CIO
CyberSaint Security

Internet of Things, for Things, and by Things

Internet of Things, for Things, and by Things

Abhik Chaudhuri

CRC Press
Taylor & Francis Group
Boca Raton London New York

CRC Press is an imprint of the
Taylor & Francis Group, an **informa** business

CRC Press
Taylor & Francis Group
6000 Broken Sound Parkway NW, Suite 300
Boca Raton, FL 33487-2742

First issued in paperback 2022

ISBN 13: 978-1-03-240182-9 (pbk)
ISBN 13: 978-1-138-71044-3 (hbk)

DOI: 10.1201/9781315200644

Library of Congress Cataloging-in-Publication Data

Names: Chaudhuri, Abhik, author.
Title: Internet of things, for things, and by things / Abhik Chaudhuri.
Description: Boca Raton, FL : CRC Press/Taylor & Francis Group, 2019. | "A CRC title, part of the Taylor & Francis imprint, a member of the Taylor & Francis Group, the academic division of T&F Informa plc." | Includes bibliographical references and index.
Identifiers: LCCN 2018017146| ISBN 9781138710443 (hbk. : acid-free paper) | ISBN 9781315200644 (ebook)
Subjects: LCSH: Internet of things.
Classification: LCC TK5105.8857 .C54 2019 | DDC 004.67/8--dc23
LC record available at https://lccn.loc.gov/2018017146

Visit the Taylor & Francis Web site at
http://www.taylorandfrancis.com

and the CRC Press Web site at
http://www.crcpress.com

Dedicated to my daughter, Sree Aishani Chaudhuri.

Contents

Foreword

The World Economic Forum's (WEF) Global Risk Report for 2018 has identified cyber attacks as the third-highest risk globally, with extreme weather events first and natural disasters second. Furthermore, data fraud/theft was listed fourth. This clearly indicates the international worries about the effect of cyberspace and the potential effect on the world. It is therefore prudent for everyone involved with cyberspace in some way, and that includes basically everyone with a smartphone, to take notice of these risks and become accustomed to them. This includes governments, private companies, non-government organizations, civil society and more. The role players within governments and private companies who should be knowledgeable about relevant cyber risks starts right at the top, with the board members and directors right down to the data-capture clerk.

At the core of the two cyberspace-related risks identified by the WEF above is of course the Internet of Things (IoT). The risks that have already been realized because of the IoT, and the risks which will be realized in the coming years, will need serious attention on all levels, as is indicated by the WEF's classification.

The Internet of Things, for Things and by Things has arrived and is here to stay. This technology development will grow to have a massive impact on all role players, as mentioned above. All such role players should holistically evaluate all the risks related to their involvement with the IoT and manage such risks as far as possible.

A big problem presently is that most role players do not have an idea of precisely what the IoT is, where it is presently used, where it can be used in the coming years and what risks will be realized by using this technology on a wider scale. Without having the knowledge and background about the IoT, it will be impossible to evaluate the relevant risks and to manage them properly. A sound foundation of knowledge is needed to guide any decision maker and user into the future, otherwise we will try to fight IoT fires all the time instead of having a well-founded IoT risk management plan for implementing and managing IoT-related systems.

This is precisely where this comprehensive and extremely useful contribution of Abhik Chaudhuri will add immense strategic value to creating the sound foundation of knowledge referred to above. This contribution can be seen as seminal in the sense that it will allow role players to get insight and knowledge about precisely what the IoT is and how it can be used, and it provides significantly more knowledge needed to get the relevant insight and background. This contribution is timely and should be a strongly advised reference work for anyone who is venturing into the risky field of the IoT.

The wide-ranging list of topics covered clearly also indicates the multi-disciplinary and multi-dimensional character of cyberspace, and specifically the IoT. This is therefore not only for technologists, as topics like privacy, ethics, education, the human dimension, governance, smart cities and regulatory aspects address many more role players than only the technologists. It therefore really addresses the multi-disciplinary character of cyberspace. By emphasizing and including these relevant chapters, Abhik Chaudhuri conveys the message that the IoT is not a technical phenomenon, but that it will have an impact on the world much larger than only technology—it can potentially change the world in ways we do not even yet understand at this point in time.

Personally, I congratulate the author in starting and eventually completing this very valuable contribution. I trust it will not be the last contribution he creates in this area, as the knowledge and experience reflected in this contribution should be leveraged by him in future—I wish him well in his future career.

I trust all readers of this book will appreciate its value in helping to improve and secure cyberspace.

Prof. Basie von Solms
Director: Centre for Cyber Security (www.cybersecurity.org.za)
Academy for Computer Science and Software Engineering
University of Johannesburg
South Africa
Associate Director: Global Cybersecurity Capacity Centre, University of Oxford, UK
Past President: IFIP

Preface

Seven hundred years ago, a single breakthrough in information technology ended one era of human history and launched another. This powerful new technology brought knowledge to the masses, which in turn leveraged a series of significant social movements, such as the Renaissance, the Reformation and the Age of Enlightenment. Those movements transformed a way of life that had existed, fundamentally unchanged, for over a millennium, and formed the basis for our modern world. That innovation was the invention of movable type.

Historically, timing is critical when you consider the impact of technology. That is because the people who are undergoing rapid societal changes have to be able to adjust to the new realities. It is important to note that the Renaissance and the Age of Enlightenment fell within a relatively short period of time. However, the type of epoch-ending technological revolution that we are currently undergoing is all taking place in a matter of twenty years, not three centuries.

That is an important distinction to keep in mind when we make our plans in the long-term. The extremely abbreviated timeframe of the current technological revolution is a critical concern because our social institutions must eventually align with our capabilities. The people who were alive in the 15th century did not live to see the 17th, which is when the eventual impacts of movable type came to fruition. So even though change was rapid on the historical scale, social institutions were able to adjust in a rational and relevant manner.

Whereas our newer and even more profound post-industrial knowledge society evolved out of the 1990s, with the advent of private networks and the commercial internet. So, this entire revolution has taken place in a matter of twenty years, not three centuries. The global networking phenomenon has evolved to include an almost ridiculously diverse set of intelligent objects; all wired together in what we have termed the "Internet of Things," or IoT.

The Internet of Things is an ecosystem. It can include any type of electronic device, which could conceivably be hooked to another device. Like every classic network environment, the IoT includes big obvious things, like computers and software systems. But, it can also encompass an unthinkable array of tiny invisible things, like sensors, and actuators and controllers. Oh my!! That is particularly true given the availability of all sorts of methods of inter-connection.

The tangled mass of potential interconnectivity can quickly become very confusing. Right now, any form of programmed logic can be embedded in any conceivable object and then be intentionally, or even unintentionally connected with a range of other components. The outcome of that is a massive collection of things that we might not necessarily recognize as related. So potentially your refrigerator and its Internet-enabled sensors might be communicating with your car's on-board systems through your home-security system which is in turn communicating with the Internet-at large, through your smart TV.

That impossible-to-comprehend diversity of interconnection is both a blessing and a curse. It is a blessing in that our social institutions can be leveraged through the immense power of network systems. That includes everything from arms of government to self-driving vehicles. The curse is the lack of any clear understanding or implicit control over what this all leads to, in terms of both the consequences, intended or otherwise, as well as the likely impacts on the fundamental safety, security and reliability of our way of life.

For instance, the October 21, 2016, Mirai DDoS attack on Dyn Systems was leveraged off of IoT botnets. The concept of an army of Internet-enabled toasters and baby monitors shutting down a major provider of DNS services is almost comical until you consider its implications and potential impacts on the underpinnings of society. So, we need to better understand what we are doing when it comes to the Internet of Things and its implications and applications in our modern world and society.

That understanding is the real value of this book. This book provides a cogent and comprehensive discussion of the real-world implications of the Internet of Things, both from the perspective of practical application, and also as it regards the management of risk and the evolution of policy to ensure the safe development and use of the IoT. The contents of this book cover the waterfront, both in terms of practical big-picture areas of consideration of the IoT, as well as the pertinent and pragmatic issues of day-to-day use of IoT devices and their potentially unintended consequences.

The knowledge contained here is something that every well-informed person in any area of corporate or governmental responsibility needs to know. And in many respects, it represents a start in the process of our ability to keep up in the race to understand and properly manage the outrageous state of technological change that we find ourselves trapped in.

Dan Shoemaker, PhD
Professor and Graduate Program Director
Center for Cybersecurity and intelligence Studies, University of Detroit Mercy
A National Security Agency Center of Academic Excellence
in Cyber Defense Education

About the Author

Abhik Chaudhuri is a Chevening Fellow (UK) and Fellow of Cloud Security Alliance (USA). He is a Domain Consultant in Cyber Security, Privacy and Policy for IT, IoT and Smart Cities, with the Design and Architecture Center of Excellence at Tata Consultancy Services (TCS). Abhik has more than 16 years of IT consulting and research experience and has been a contributing member of NIST CPS PWG. As Co-Editor of ISO/IEC JTC1/SC27 Abhik is providing thought leadership in developing global cyber security and privacy standards. He is a Corporate Member of Cloud Security Alliance's International Standardization Council and member of Editorial Boards of EDPACS Journal (Taylor & Francis, USA), the Journal of Data Protection and Privacy (Henry Stewart Publications, UK) and IEEE Ethics and Policy in Technology eNewsletters. Abhik has multiple cyber thought leadership contributions in leading international journals, books, Cyber Security and Privacy Frameworks, and Digital Policy. He is a member of the UN's Internet Governance Forum, IIC, ISOC, IAPP, IEEE's IoT Technical Community, IEEE Global Initiative for Ethical Considerations in Artificial Intelligence and Autonomous Systems, and IEEE Standards Association.

Words from the Author

Worldwide, the concept of the Internet of Things (IoT) is catching up with digital equipment manufacturers, businesses and users. There is a growing trend of interconnecting humans with digital devices for a smarter living experience.

This book is influenced by Abraham Lincoln's famous quote "Government of the people, by the people, for the people, shall not perish from the Earth." The Internet as a technology has become a basic necessity in today's world. This necessity is being further aggravated with the emergence of IoT technology that is promising to provide new digital business models and services by utilizing perception and analytics of contextual data. The availability of cheap computing devices and the capability to analyze huge volume of data gathered by IoT devices located in diverse environments is making it possible to create unprecedented service opportunities across business domains.

Critically, this book aims to help businesses, policy makers, technologists, educators, regulators and netizens across the globe to understand IoT technology; its potential applications; its security and privacy aspects; its key necessities like governance, risk management and regulatory compliance and the philosophical aspects of this technology that are necessary to support an ethical, safe and secure digitally enhanced environment in which people can live smarter.

This book is divided into three sections, comprising 11 chapters that divide the IoT concept into three key areas:

Section I—'of Things'
Section II—'for Things'
Section III—'by Things'

Section I describes the inherent technology of the IoT, the architectural components and the philosophy behind this emerging technology. This section is about the 'of Things' part of the IoT.

Section II shows the various potential applications of the IoT that can bring benefits to human society. It includes a glimpse of available IoT products/services and smart-city designs as well as futuristic, inclusive designs that are being explored by innovation labs and organizations. This section is about the 'for Things' part of the IoT.

Section III discusses various necessities to provide a secured and trustworthy IoT service. It includes chapters on IoT security; IoT privacy; IoT GRC, standards and regulatory needs; the importance of the assurance function for trustworthy IoT services; the role of net neutrality in the IoT and the human dimension of the IoT for creating a smart 'white box' society. This section is about the 'by Things' part of the IoT.

The 12th chapter is on the novel 'IoT Privacy Framework' and 'IoT Privacy by Design Principles' co-developed with Dr. Ann Cavoukian that has been added in Appendix 1 for the benefit of the readers. Appendix 2 provides a list of information risk assessment frameworks and standards for digital services, and Appendix 3 provides a glimpse of various global initiatives related to the IoT and smart cities.

This book can be a knowledge guide for all kinds of readers that include business leaders, technologists, regulators, netizens, digital enthusiasts and auditors who are interested in knowing about the key aspects of IoT technology and smart cities. It is also for smart city councils and governments across the globe that are exploring or implementing IoT-enabled smart services.

I hope my sincere effort will help readers understand the technology, philosophy, potentialities and challenges of the IoT and realize the benefits of this emerging technology.

Abhik Chaudhuri
Chevening Fellow (UK)—Cyber Security, Privacy & Policy
Fellow of Cloud Security Alliance (USA)
Co-Editor: ISO/IEC JTC1 SC27

Acknowledgments

I am thankful to Dan Swanson for his continuous support and encouragement to write this book. Special thanks to Rich and his publication team.

I have received immense support and encouragement from Professor Basie von Solms and Professor Dan Shoemaker. I am thankful to them for sharing their insights in the Foreword and Preface for this book.

I am thankful to the thought leaders Dr. Ann Cavoukian, Professor Harold "Bud" Lawson, James F. Moore, Professor Irving Wladawsky-Berger, Peter T. Lewis, Dr. Maria Bada, Douglas Laney, Ian Abbott-Donnelly, Florence Hudson, Professor Joy Sen, Mark J. Toomey and Scott M. Schlimmer for their support and for sharing their messages in the 'Messages from Global Leaders' pages of this book.

Special Thanks to David Gascón, Michelle Chibba, Chuck Benson and Dr. Luca Belli for sharing their thoughts in the 'Five Questions to the Global Leader' sections in this book.

I would like to thank my seniors and colleagues at my workplace for their support and encouragement.

I thank my colleagues, co-editors, friends and thought leaders at Cloud Security Aliance (CSA), CSA International Standardization Council, ISO, UN IGF, ISOC, IETF, NIST CPS PWG, IEEE, ISACA, IAPP and in other institutions, academia and industries whom I have worked or interacted with in various efforts.

This work would not have materialized without the support I received from my family. I am truly indebted to my parents, spouse and daughter for their continuous encouragement in writing this book.

Abhik Chaudhuri
Chevening Fellow (UK)—Cyber Security, Privacy & Policy
Fellow of Cloud Security Alliance (USA)
Co-Editor: ISO/IEC JTC1 SC27

Acknowledgments

'OF THINGS'

I

Chapter 1

Internet of Things and Its Potential

The Internet of Things has the potential to change the world, just as the Internet did. Maybe even more so.

Kevin Ashton

After reading this chapter you will be able to:

- Define Internet of Things
- Know the projected global growth of Internet of Things
- Understand the importance of Internet of Things in potential applications across various businesses and service domains
- Understand the aspects of the Industrial Internet of Things
- Understand the relationship between Cyber Physical Systems and the Internet of Things
- Understand the complex ecosystem of the Internet of Things
- Interpret the role of the Interet of Things in realizing the United Nations' 2030 Agenda for Sustainable Development.

The Internet of Things: Definition

The 'Internet of Things' (IoT) is an emerging technology that enables interaction of uniquely identifiable computing devices that can be embedded with other interfaces like machines and humans, linked via wired and wireless networks, to capture contextual data from the environment it has been exposed to and create an information network to provide new functionalities and digital business models. It is also popularly referred to by the abbreviated name of 'IoT'.

In the September 1985 Congressional Black Caucus 15th Legislative Conference at the Washington Hilton Hotel, Peter T. Lewis delivered a talk focusing on 'cellular' in which he used the term 'Internet of Things' in the context of describing machine and device wireless connectivity as a means to create additional revenue centers for cellular networks through such connectivity. In this speech, the FCC representative, Lewis had mentioned that:

> By connecting devices such as traffic signal control boxes, underground gas station tanks and home refrigerators to supervisory control systems, modems, auto-dialers and cellular phones, we can transmit status of these devices to cell sites, then pipe that data through the Internet and address it to people near and far that need that information. I predict that not only humans, but machines and other things will interactively communicate via the Internet. The Internet of Things, or IoT, is the integration of people, processes and technology with connectable devices and sensors to enable remote monitoring, status, manipulation and evaluation of trends of such devices. When all these technologies and voluminous amounts of Things are interfaced together—namely, devices/machines, supervisory controllers, cellular and the Internet, there is nothing we cannot connect to and communicate with. What I am calling the Internet of Things will be far reaching.[*][†]

Kevin Ashton, the co-founder of the Auto-ID Center at the Massachusetts Institute of Technology, had provided a glimpse of this emerging technology in one of his articles, where he mentioned:

> Today computers—and, therefore, the Internet—are almost wholly dependent on human beings for information. Nearly all of the roughly 50 petabytes (a petabyte is 1,024 terabytes) of data available on the Internet were first captured and created by human beings—by typing, pressing a record button, taking a digital picture or scanning a bar code. Conventional diagrams of the Internet include servers and routers and so on, but they leave out the most numerous and important routers of all: people. The problem is, people have limited time, attention and accuracy—all of which means they are not very good at capturing data

---

[*] Sharma, C. (n.d.). Correcting the IoT History. Retrieved from http://www.chetansharma .com/correcting-the-iot-history/

[†] Still, L., & Carter, A. (1985, November). National Roundup. The Afro-American. Baltimore, Md.: Afro-American Co. Retrieved from https://news.google.com/newspapers?nid=2211&dat =19851109&id=TyImAAAAIBAJ&sjid=%20Tf4FAAAAIBAJ&pg=2555,1699142&hl=en

about things in the real world. . . . We need to empower computers with their own means of gathering information, so they can see, hear and smell the world for themselves, in all its random glory.*

In this decade with the advancement in computing, sensing and storage technologies, we are gradually realizing the era of the IoT. IoT technology is gaining popularity across the globe to provide new digital business models and services by utilizing the capability of sensors to capture data from different sources that was not possible even a decade ago. The availability of cheap computing devices and the processing power of information from contextual data gathered on a continuous basis by IoT devices located in disperse environments and machines is making it possible to create new opportunities in services and solutions across business domains.

The IoT, being an in-focus emerging technology, does not yet have a common definition across industry, academia, research and standards bodies. However, the various efforts of defining the IoT have mostly imbibed a common theme of triggering defined beneficial action that is based on continuous analysis of contextual data, as depicted in the following paragraphs.

According to Cisco, the IoT is the 'Internet of Everything', which "links objects to the Internet, enabling data and insights never available before,"[†] and it predicts that "500 billion devices are expected to be connected to the Internet by 2030."[‡]

The International Telecommunication Union (ITU) has defined the IoT in Recommendation ITU-T Y.2060 (06/2012) as "a global infrastructure for the information society, enabling advanced services by interconnecting (physical and virtual) things based on existing and evolving interoperable information and communication technologies."[§]

The European Commission's Digital Agenda for Europe defines the IoT as a "technology and a market development based on the inter-connection of everyday objects among themselves and applications. IoT will enable an ecosystem of smart applications and services which will improve and simplify"[¶] human lives.

McKinsey & Company has defined the IoT as "sensors and actuators connected by networks to computing systems. These systems can monitor or manage

---

* Ashton, K. (2009). That 'internet of things' thing. RFID journal, 22(7), 97–114. Retrieved from http://www.rfidjournal.com/articles/view?4986
† Cisco. (n.d.). Internet of Things. Retrieved from https://www.cisco.com/c/en_in/solutions /internet-of-things/overview.html
‡ Cisco. (2016). Internet of Things At-a-Glance. Retrieved from https://www.cisco.com/c/dam /en/us/products/collateral/se/internet-of-things/at-a-glance-c45-731471.pdf
§ International Telecommunications Union. (2015). Overview of the Internet of things. Retrieved from http://handle.itu.int/11.1002/1000/11559
¶ European Commission. (n.d.). The Internet of things. Retrieved from http://ec.europa.eu /digital-agenda/en/internet-things

the health and actions of connected objects and machines. Connected sensors can also monitor the natural world, people, and animals."*

According to Goldman Sachs, the IoT "connects devices such as everyday consumer objects and industrial equipment onto the network, enabling information gathering and management of these devices via software to increase efficiency, enable new services, or achieve other health, safety or environmental benefits."† The IoT is emerging as the third wave of Internet after the fixed internet wave of the 1990s and the mobile wave of the 2000s.

The five key attributes based on the 'S-E-N-S-E' framework‡ (Goldman Sachs) that distinguish the IoT from the regular form of Internet are:

1. **S**ensing - Leveraging sensors to generate contextual data
2. **E**fficient - Enhances efficiency in productivity terms by adding intelligence to 'things'
3. **N**etworked - Creates a network of 'things'
4. **S**pecialized - Ultimate use is focused on domain-based specialized offerings
5. **E**verywhere - Can be deployed everywhere for a ubiquitous presence as per objectivity of use.

According to the National Institute of Standards and Technology (NIST), sensors are one of the five core system primitives that "form the basic building blocks for a Network of 'Things' (NoT), including the Internet of Things (IoT)."§

The Internet Society refers to the IoT as "scenarios where network connectivity and computing capability extends to objects, sensors and everyday items not normally considered computers, allowing these devices to generate, exchange and consume data with minimal human intervention. There is, however, no single, universal definition."¶

The Internet Engineering Task Force claims that the IoT "are devices that are (1) designed to be dependent on the Internet, where such a device would not have

---

* Löffler, M., Münstermann, B., Schumacher, T., Mokwa, C., & Behm, S. (2016, August). Insurers need to plug into the Internet of Things – or risk falling behind. McKinsey & Company. Retrieved from https://www.mckinsey.com/~/media/McKinsey/Industries/Financial Services /Our Insights/European insurance practice report on Internet of Things/McKinsey-Insurers need to plug into the Internet of Things or risk falling behind.ashx
† Goldman Sachs. (2014, September). The Internet of Things: Making sense of the next megatrend. Retrieved from http://www.goldmansachs.com/our-thinking/outlook/internet-of-things /iot-report.pdf
‡ *Ibid*
§ NIST. (2016, July). NIST's Network-of-Things Model Builds Foundation to Help Define the Internet of Things. Retrieved from http://nist.gov/itl/csd/nists-network-of-things-model -builds-foundation-to-help-define-the-internet-of-things.cfm
¶ Rose, K., Eldridge, S., & Chapin, L. (2015, October). The Internet of Things: An Overview. ISOC. Retrieved from https://www.internetsociety.org/wp-content/uploads/2017/08/ISOC -IoT-Overview-20151221-en.pdf

depended on the Internet previously, and (2) rapidly manufactured, homogeneously configured, and deployed across the Internet."*

A.T. Kearney defines the IoT as a "seamless combination of embedded intelligence, ubiquitous connectivity, and deep analytical insights that creates unique and disruptive value for companies, individuals, and societies."†

Intel defines the IoT as "devices that are connecting to the internet, integrating greater compute capabilities, and using data analytics to extract valuable information."‡

According to Gartner, the IoT is "the network of physical objects that contain embedded technology to communicate and sense or interact with their internal states or the external environment."§

The IoT is also referred to as the 'Industrial Internet of Things', or 'Industry 4.0'. Microsoft refers to the IoT as the 'Internet of Your Thing'.¶

The Hype Around the Internet of Things

As per the 'Hype Cycle for Emerging Technologies from Gartner',** as in Figure 1.1, the IoT reached the 'Peak of Inflated Expectations' in 2015. IoT adoption is on the cusp of a multi-year, multi-fold growth. Gartner has predicted that IoT sensors, devices and systems will grow to 25 billion units by 2020.††

This growth prospect is fueled by continuous reduction of the cost of computing power and the adoption of IPV6 technology, which allows connecting almost everything on earth with unique IP addresses (extension from 32-bit IP address to 128-bit address space). As we move from IPV4 to IPV6, we will be able to assign unique IP addresses to each and every device on earth. This is showing tremendous opportunity to interconnect existing and new services in utilities, business, healthcare, education and other services over the Internet to create smart-living offerings in cities with real-time data analytics and decision-support systems.

---

* Plonka, D. (2016, November). The Internet of Things Unchecked. IETF. Retrieved from https://www.ietfjournal.org/the-internet-of-things-unchecked/
† A.T. Kearney Inc. (2016). The Internet of Things: A New Path to European Prosperity. Retrieved from https://www.atkearney.com/documents/10192/7125406/The+Internet+of+Things-A+New +Path+to+European+Prosperity.pdf
‡ Intel Corporation. (2014). Intel and the Internet of Things. Retrieved from https://newsroom .intel.com/press-kits/intel-and-the-internet-of-things-2/
§ Gartner Inc. (n.d.). IT Glossary. Retrieved from http://www.gartner.com/it-glossary/internet -of-things/
¶ Microsoft Corporation. (n.d.). Internet of Things. Retrieved from https://www.microsoft .com/en-in/internet-of-things/
** Gartner Inc. (2015). Hype cycle for emerging technologies. Retrieved from http://www.gartner .com/document/3100227?ref=lib
†† Gartner Inc. (2014). Gartner says 4.9 billion connected "things" will be in use in 2015. Retrieved from http://www.gartner.com/newsroom/id/2905717

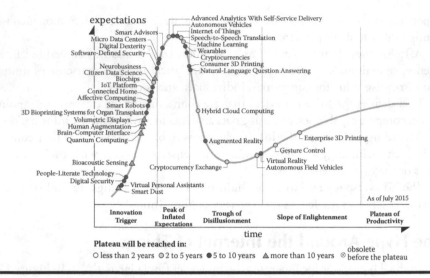

Figure 1.1 Gartner's Hype Cycle for Emerging Technologies.

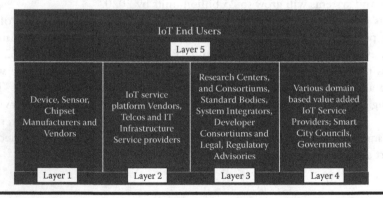

Figure 1.2 The IoT ecosystem.

The Internet of Things—A Complex Ecosystem

The IoT has many players across various layers of its evolutionary ecosystem, as depicted in Figure 1.2. If we segregate these players based on their functionality in the ecosystem, then we will have the manufacturers of IoT devices, sensors, chipsets and the vendors in layer 1. Above this layer we have layer 2 for Telecom service providers, IoT platform vendors and the IT infrastructure service providers. Layer 3 will have research consortiums, standards bodies, system integrators, developer consortiums and legal, regulatory advisors. Layer 4 will have Smart-city councils, governments and various domain-based, value-added IoT service providers. Layer 5 has end users of various IoT products and services, who are experiencing the

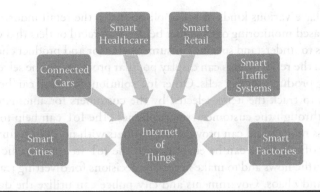

Figure 1.3   Potential smart services with the IoT.

benefits of the IoT individually, in groups and as a society. As each of these layers build upon the other to provide a safe, secured and trustworthy IoT service, it is the end users who become the common binder for this layered ecosystem through their usage, concerns, feedback and mass adoption.

Potential Applications of the Internet of Things

As shown in Figure 1.3 below, connected cars, smart healthcare, smart retail, smart traffic systems, smart factories and smart cities are all great examples of the potential of the IoT that we can design and implement for our benefits.

With the IoT we can design connected cars that capture contextual data from the fuel tank and combine it with the average driving trend to provide smart indicators that can inform us in advance about predictive fuel usage and when to refuel the vehicle. IoT-enabled health checks of the automotive components, GRPS-based location monitoring, crash prevention with car-to-car smart communication on busy highways, travel-route optimization with real-time traffic analysis and driverless communication are some of the promising features of IoT-enabled cars.

We have smart watches today that help to monitor and report the physical activities of humans as well as critical health parameters like blood pressure, heart rate, etc. Remote patient monitoring with IoT-enabled wearable medical devices is a promising digital service that can revolutionize the healthcare industry. For example, patients with heart ailments, diabetes and kidney problems can be remotely monitored by doctors from data generated by the IoT-enabled wearable or implanted devices on such patients for decision-based health data analysis and medication. With the IoT, specific healthcare solutions can be provided like round-the-clock blood glucose level monitoring and automated injection of insulin in a patient's blood by insulin pumps implanted in the patient's body, based on variations of glucose levels remotely monitored by doctors.

We can have various kinds of IoT applications in the retail industry like wireless sensor-based monitoring of customer behavior in retail outlets that can help the retail owners to understand specific consumer behavior and product choices. Based on these data the retail owners can display popular products on the selves, replacing slow-moving products for fast sells. Other IoT solutions for retail can be smart carting facilities to track the items selected by the customers for automated payment and billing through the customers' smart phones. The IoT can help to implement such business models and can provide retail outlets with no POS counters.

Smart traffic systems can be enabled with the IoT in a city traffic network to determine traffic flows and to make necessary decisions for diverting traffic to avoid congestion and chaos. Governments and city police can utilize the data gathered over specific time periods from smart traffic systems to decide on infrastructure requirements for smooth traffic movements across the city network.

Factories can be made more energy efficient and the supply chain can be made more integrated with demand and market needs by using the data generated by IoT devices fitted to individual components of the machines on the floor. Data gathered from these IoT devices can be analytically processed to provide just-in-time information to the engineers and managers for prompt decision making and efficiency in material usage, power utilization and effluent treatment.

The benefits of the IoT can be extended to smart city services where various IoT-enabled smart services will interconnect to create interdependent systems of systems. IoT-enabled smart cities are described in detail in the following section.

Smart cities will have IoT-enabled services like water management and distribution, energy generation and transmission, logistics, garbage disposal and management, smart street-lighting systems, citizen-connected healthcare, smart surveillance and security systems, e-governance systems and other smart services.

The IoT for The United Nations' 2030 Agenda for Sustainable Development

The IoT's potential can be utilized to realize the Sustainable Development Goals (SDGs) of the United Nations' "2030 Agenda for Sustainable Development."[*] Paragraph 15 of the Agenda, which aims to transform our world, specifically mentions and acknowledges the potential of information and communication technologies for development. It says: "The spread of information and communications technology and global interconnectedness has great potential to accelerate human progress, to bridge the digital divide and to develop knowledge societies, as does scientific and technological innovation across areas as diverse as medicine and energy."[†]

---

[*] United Nations. (2015). Transforming our world: the 2030 Agenda for Sustainable Development. Retrieved from https://sustainabledevelopment.un.org/content/documents /21252030%20Agenda%20for%20Sustainable%20Development%20web.pdf
[†] Ibid.

Table 1.1 IoT Interventions Mapped to the Un's Millenium Development Goals (MDGs) and Sustainable Development Goals (SDGs)*

Sector	MDG	SDG	Examples
Health, water and sanitation	MDG 4: Child health MDG 5: Maternal health MDG 6: Combat HIV/AIDS, malaria and other diseases	SDG 3: Ensure healthy lives and promote well-being for all at all ages. SDG 6: Ensure availability and sustainable management of water and sanitation for all.	Sensor- and SMS-enabled village water pumps (Rwanda, Kenya); GSM-connected refrigeration for vaccine delivery in the 'cold chain' (global); sensor-enabled 'band aids' to monitor Ebola patients' ECG, heart rate, oxygen saturation, body temperature, respiratory rate and position, all remotely (West Africa); water stream gauge with sonar range sensor to monitor river flow and depth (Honduras); water flow sensors and motion detectors in latrines to monitor efficacy of hygiene training and intervention (Indonesia).
Agriculture and livelihoods	MDG 1: End Poverty and hunger	SDG 1: End poverty in all its forms everywhere. SDG 8: Promote sustained, inclusive and sustainable economic growth, full and productive employment and decent work for all. SDG 2: End hunger, achieve food security and improve nutrition, and promote sustainable agriculture.	Connected micro-weather stations improving localized weather data and provision of crop-failure insurance (Kenya); low-cost mobile-controlled micro-irrigation pumps (India); soil-monitoring sensors used to improve tea plantation production (Sri Lanka, Rwanda); RFID-based food supply testing and tracking system (India) and RFID-based livestock programmes for tracking, theft prevention and vaccination records (Botswana, Senegal and Namibia).

(Continued)

Table 1.1 (Continued) IoT Interventions Mapped to the Un's Millenium Development Goals (MDGs) and Sustainable Development Goals (SDGs)*

Sector	MDG	SDG	Examples
Education	MDG 2: Universal education	SDG 4: Ensure inclusive and equitable quality education and promote lifelong learning opportunities for all.	Smart identity cards with biometric features for all public school students to improve service delivery (Nigeria); biometric clocking device to improve teacher attendance in real-time (South Africa).
The environment and conservation	MDG 7: Environment	SDG 12: Ensure sustainable consumption and production patterns. SDG 13: Take urgent action to combat climate change and its impacts. SDG 14: Conserve and sustainably use the oceans, seas and marine resources for sustainable development. SDG 15: Protect, restore and promote sustainable use of terrestrial ecosystems, sustainably manage forests, combat desertification, halt and reverse land degradation, and halt biodiversity loss.	Radio-based cloud-connected devices to identify and track the presence of illegal fishermen (Timor-Leste); air pollution sensors to monitor urban outdoor air pollution (Benin); acoustic sensors to monitor sea bird populations (global); sensors and connectivity to protect game park perimeters and track animals (Africa); connected unmanned aerial vehicles to monitor national parks and connect images from camera traps (UAE); acoustic sensors in tropical rainforests to 'listen' for illegal logging (Indonesia).

(Continued)

Table 1.1 (Continued) IoT Interventions Mapped to the Un's Millenium Development Goals (MDGs) and Sustainable Development Goals (SDGs)*

Sector	MDG	SDG	Examples
Resiliency, infrastructure and energy		SDG 7: Ensure access to affordable, reliable, sustainable and modern energy for all. SDG 9: Build resilient infrastructure, promote inclusive and sustainable industrialization, and foster innovation. SDG 11: Make cities and human settlements inclusive, safe, resilient and sustainable.	Networked fire/smoke alarms in high-density urban slums/informal settlements (Kenya, South Africa); Connected buoys as part of the tsunami monitoring system (Indian Ocean); off-grid micro-solar electricity systems for lower-income households (East Africa, India); connected black carbon and use sensors to monitor cook stoves (Sudan); sensor-connected matatus (mini-buses) to track velocity, acceleration and braking to curb the dangerous operation of public transportation (Kenya).
Governance and human rights		SDG 10: Reduce inequality within and among countries. SDG 16: Promote peaceful and inclusive societies for sustainable development, provide access to justice for all and build effective, accountable and inclusive institutions at all levels.	Retinal scans used for ATMs to provide secure biometric cash assistance to displaced refugees (Jordan).

* Biggs, P., Garrity, J., LaSalle, C., Polomska, A., & Pepper, R. (2016). Harnessing the Internet of Things for global development. International Telecommunication Union. Retrieved from https://www.sbs.ox.ac.uk/cybersecurity-capacity/system/files/Harnessing-IoT-Global-Development.pdf

The report "Harnessing the Internet of Things for Global Development," written as a contribution to the ITU/UNESCO Broadband Commmission for Sustainable Development, provides multiple examples of IoT interventions that have been mapped to the Millennium Development Goals (MDGs) and Sustainable Development Goals (SDGs) in various sectors like:

■ Health, water and sanitation
■ Agriculture and livelihoods
■ Education
■ The environment and conservation
■ Resiliency, infrastructure and energy
■ Governance and human rights.

These mappings are derived from the report and depicted in Table 1.1.

Apart from the above-mentioned possibilities, a gamut of smart services can be designed across domains by harnessing the potential of IoT technology. Further details on the potential application of IoT technology have been provided in Section II of this book.

Suggested Reading

Atzori, L., Iera, A., & Morabito, G. (2010). The internet of things: A survey. *Computer Networks, 54*(15), 2787–2805.

Bandyopadhyay, D., & Sen, J. (2011). Internet of things: Applications and challenges in technology and standardization. *Wireless Personal Communications, 58*(1), 49–69.

Benardos, P. G., & Vosniakos, G. C. (2017). Internet of things and industrial applications for precision machining. In *Solid state phenomena* (Vol. 261, pp. 440–447). Trans Tech Publications.

Bo, Y., & Wang, H. (2011, May). The application of cloud computing and the internet of things in agriculture and forestry. In *International Joint Conference on Service Sciences (IJCSS), 2011* (pp. 168–172). IEEE.

Bughin, J., Chui, M., and Manyika J. (2015, August). An executive's guide to the internet of things. Retrieved from http://www.mckinsey.com/business-functions/business-technology/our-insights/an-executives-guide-to-the-internet-of-things

Bui, N., & Zorzi, M. (2011, October). Health care applications: A solution based on the internet of things. In *Proceedings of the 4th International Symposium on Applied Sciences in Biomedical and Communication Technologies* (p. 131). ACM.

Cui, X. (2016). The internet of things. In *Ethical ripples of creativity and innovation* (pp. 61–68). Palgrave Macmillan, London.

Da Xu, L., He, W., & Li, S. (2014). Internet of things in industries: A survey. *IEEE Transactions on Industrial Informatics, 10*(4), 2233–2243.

Dohr, A., Modre-Opsrian, R., Drobics, M., Hayn, D., & Schreier, G. (2010, April). The internet of things for ambient assisted living. In *Seventh International Conference on Information Technology: New Generations (ITNG), 2010* (pp. 804–809). IEEE

European Research Cluster on the Internet of Things. (n.d.). Internet of things. Retrieved from http://www.internet-of-things-research.eu/about_iot.htm

Fleisch, E. (2010). What is the Internet of things? An economic perspective. *Economics, Management & Financial Markets, 5*(2).

Gubbi, J., Buyya, R., Marusic, S., & Palaniswami, M. (2013). Internet of things (iot): A vision, architectural elements, and future directions. *Future Generation Computer systems, 29*(7), 1645–1660.

IEEE IoT Toolkit. (n.d.). Retrieved from https://iot.ieee.org/about/ieee-iot-toolkit.html

Jeschke, S., Brecher, C., Meisen, T., Özdemir, D., & Eschert, T. (2017). Industrial internet of things and cyber manufacturing systems. In *Industrial Internet of Things* (pp. 3–19). Springer International Publishing.

Kopetz, H. (2011). Internet of things. In *Real-time systems* (pp. 307–323). Springer US.

Mainetti, L., Patrono, L., & Vilei, A. (2011, September). Evolution of wireless sensor networks towards the internet of things: A survey. In *19th International Conference on Software, Telecommunications and Computer Networks (SoftCOM), 2011* (pp. 1–6). IEEE.

Miorandi, D., Sicari, S., De Pellegrini, F., & Chlamtac, I. (2012). Internet of things: Vision, applications and research challenges. *Ad Hoc Networks, 10*(7), 1497–1516.

Neagle, C. (2014, July). A guide to the confusing internet of things standards world. Retrieved from http://www.networkworld.com/article/2456421/internet-of-things/a -guide-to-the-confusing-internet-of-things-standards-world.html

Oberländer, A. M., Röglinger, M., Rosemann, M., & Kees, A. (2017). Conceptualizing business-to-thing interactions—A sociomaterial perspective on the Internet of Things. *European Journal of Information Systems*, 1–17.

Perera, C., Zaslavsky, A., Christen, P., & Georgakopoulos, D. (2014). Sensing as a service model for smart cities supported by internet of things. *Transactions on Emerging Telecommunications Technologies, 25*(1), 81–93.

Perera, C., Zaslavsky, A., Christen, P., & Georgakopoulos, D. (2014). Context aware computing for the internet of things: A survey. *IEEE Communications Surveys & Tutorials, 16*(1), 414–454.

Scott, A. (n.d.). 8 ways the internet of things will change the way we live and work. Retrieved from https://www.theglobeandmail.com/report-on-business/rob-magazine /the-future-is-smart/article24586994/

Spanò, E., Niccolini, L., Di Pascoli, S., & Iannacconeluca, G. (2015). Last-meter smart grid embedded in an internet-of-things platform. *IEEE Transactions on Smart Grid, 6*(1), 468–476.

Sun, Q. B., Liu, J., Li, S., Fan, C. X., & Sun, J. J. (2010). Internet of things: Summarize on concepts, architecture and key technology problem [J]. *Journal of Beijing University of Posts and Telecommunications, 3*(3), 1–9.

Sundmaeker, H., Guillemin, P., Friess, P., & Woelfflé, S. (2010). Vision and challenges for realising the internet of things. *Cluster of European Research Projects on the Internet of Things, European Commision, 3*(3), 34–36.

Tan, L., & Wang, N. (2010, August). Future internet: The internet of things. In *3rd International Conference on Advanced Computer Theory and Engineering (ICACTE), 2010* (Vol. 5, pp. V5–376). IEEE.

McKinsey Global Institute (2015, August). The internet of things: Five critical questions. Retrieved from http://www.mckinsey.com/industries/high-tech/our-insights/the-internet -of-things-five-critical-questions

Vlacheas, P., Giaffreda, R., Stavroulaki, V., Kelaidonis, D., Foteinos, V., Poulios, G., ... & Moessner, K. (2013). Enabling smart cities through a cognitive management framework for the internet of things. *IEEE communications magazine, 51*(6), 102–111.

Watson. (n.d.). Internet of things. Retrieved from http://www.ibm.com/internet-of-things/

Wortmann, F., & Flüchter, K. (2015). Internet of things. *Business & Information Systems Engineering, 57*(3), 221–224.

Xia, F., Yang, L. T., Wang, L., & Vinel, A. (2012). Internet of things. *International Journal of Communication Systems, 25*(9), 1101.

Yun, M., & Yuxin, B. (2010, June). Research on the architecture and key technology of Internet of Things (IoT) applied on smart grid. In *International Conference on Advances in Energy Engineering (ICAEE), 2010* (pp. 69–72). IEEE.

Zanella, A., Bui, N., Castellani, A., Vangelista, L., & Zorzi, M. (2014). Internet of things for smart cities. *IEEE Internet of Things Journal, 1*(1), 22–32.

Chapter 2

IoT Architecture

Architecture is ... a rational procedure to do sensible and hopefully beautiful things.

Harry Seidler

After reading this chapter you will be able to:

- Understand the system design of IoT technology at the component level
- Understand the enterprise and security architecture of IoT technology
- Gain an insight on edge or fog computing
- Understand the system architecture of the Industrial IoT (IIoT)
- Understand the relevance of middleware architecture in IoT
- Gain an insight on cyber physical systems (CPS).

Introduction

According to ISO/IEC/IEEE 42010:2011, the architecture of a system is defined as "fundamental concepts or properties of a system in its environment embodied in its elements, relationships, and in the principles of its design and evolution."*

An Internet of Things architecture strategy is crucial to realize the potential of this emerging technology. An event-driven architecture for IoT implementation with security features is necessary to build confidence in IoT services, considering the scale and complexity of such implementation, the volume of contextual data generated from the environment, and the multiple stakeholders of the IoT ecosystem.

---

* ISO/IEC/IEEE 42010:2011 Systems and software engineering—Architecture description. (2011–2012). Retrieved from https://www.iso.org/standard/50508.html

According to Gartner,* security for IoT is a fundamental architectural building block, so emerging IoT security scenarios should be adapted in the IoT architecture.

Key Components of IoT Architecture

The key components of IoT architecture are sensors, actuators, data storage, control systems, a digital communication medium and an application for information-based action. These components are discussed here in detail.

Sensors and Actuators

A sensor has been defined by the IEEE as "an electronic device that produces electrical, optical, or digital data derived from a physical condition or event. Data produced from sensors is then electronically transferred, by another device, into information (output) that is useful in decision making done by 'intelligent' devices or individuals (people)."[†]

An actuator is defined as "a mechanical device that accepts a data signal and performs an action based on that signal."[‡]

Sensors are precisely designed as per the requirement of their sensing function, like temperature sensors, pressure sensors, proximity sensors, infrared sensors, water quality sensors, smoke sensors, chemical sensors and so on.

Libelium, the leader in wireless sensing, has developed a variety of sensors for IoT applications and smart city functions. They have published a compilation of 50 cutting edge IoT sensor applications for a smarter world that:

> is grouped in 12 different verticals, showing how the Internet of Things is becoming the next technological revolution. It includes the most trendy scenarios, like Smart Cities where sensors can offer us services like Smart Parking—to find free parking spots in the streets—or managing the intensity of the luminosity in street lights to save energy. Climate change, environmental protection, water quality or CO_2 emissions are also addressed by sensor networks....[§]

The sensors specifically designed for smart cities are depicted in Table 2.1 below, which provides a glimpse of the potential of IoT technology to make cities smarter and better.

---

* Gartner. (2017). Internet of things—architecture remains a core opportunity and challenge: A Gartner trend insight report. Retrieved from https://www.gartner.com/doc/3586317?ref=Site Search&sthkw=Adaptive%20Security&fnl=search&srcId=1-3478922254
† IEEE. (n.d.). Retrieved from http://grouper.ieee.org/groups/1451/6/TermsDefinitions.htm#sensor
‡ Ibid.
§ 50 sensor applications for a smarter world. Get inspired! (2012). Retrieved from http://www .libelium.com/50_sensor_applications/

Table 2.1 Sensors for Smart City Applications*

#	*Sensors for Smart Cities*	*Application*
1	Smart parking	Monitoring of parking spaces availability in the city.
2	Structural health	Monitoring of vibrations and material conditions in buildings, bridges and historical monuments.
3	Urban noise maps	Sound monitoring in bar areas and centric zones in real time.
4	Smartphone detection	Detect iPhone and Android devices and in general any device which works with WiFi or Bluetooth interfaces.
5	Electromagnetic field levels	Measurement of the energy radiated by cell stations and WiFi routers.
6	Traffic congestion	Monitoring of vehicles and pedestrian levels to optimize driving and walking routes.
7	Smart lighting	Intelligent and weather-adaptive lighting in street lights.
8	Waste management	Detection of rubbish levels in containers to optimize trash collection routes.
9	Smart roads	Intelligent highways with warning messages and detours according to climate conditions and unexpected events like accidents or traffic jams.

* 50 sensor applications for a smarter world. Get inspired! (2012). Retrieved from http://www.libelium.com/50_sensor_applications/.

FIVE QUESTIONS TO THE GLOBAL LEADER

David Gascón
Libelium's co-founder and CTO

1. What is Libelium's vision for a smart world?

   Libelium sensors are helping to prevent pests and improve productivity in agricultural crops, are also helping to reduce pollution and traffic in several smart cities, have been involved in projects to predict volcanic activity, are trying to universalize access to a health system to populations far from hospitals, and have even arrived in space onboard a satellite to study solar storms. Libelium's wireless sensor platform is present in projects of great economic and social impact to address the great challenges humanity faces in this century. So, we are proud to affirm that our technology leaves its mark on the world, improving people's quality of life. In fact, Libelium designs and manufactures hardware to deploy wireless sensor networks to create fast and reliable solutions for the smart cities and the IoT market. Our smart sensor technology allows us to monitor any object or environment and send this information in real time wirelessly to the Internet. It means that it allows 'Things' to communicate with the Internet. This achieves greater efficiency and optimization of any process as well as decision making based on real data. So the reach of its improvements is infinite. The Internet of Things will affect all markets: cities, logistics, security, agriculture, health, home automation, energy…. Any process can be optimized if its variables can be measured, analyzed and used in decision-making processes. The opportunities are endless, it takes a whole legion of companies and entrepreneurs that are dedicated to create sensors, install them and create applications for a smarter world.

2. What are the most popular IoT sensors from Libelium's portfolio?

   Libelium's sensor platform is open source, horizontal and scalable. This allows other products to be developed based on our platform without having to pay for patents or trademarks. The platform has been designed to implement solutions for agriculture, water management, smart cities or any other sector as it can integrate more than 120 types of sensors. Thanks to the horizontal nature and scalability of Libelium's IoT platform, any proof of concept or small project can be expanded and modified over time without having to carry out changes in the installation.

The main sectors and applications of Libelium technology are the following:

■ Smart management of car parks in cities: free space detection and wireless communication to provide information to the user.

■ Monitoring of noise and air pollution levels in cities or other environments to establish alerts, reduce emissions and develop public policies according to reality. It provides more accuracy in monitoring gas concentration levels (CO, NO2, O3 and SO2) and the amount of PM1, PM2.5 and PM10 dust particles, which allows cities and government agencies to meet international air quality index (AQI) compliance values, and better advise the public about air quality and health risks.

■ Control of water quality in rivers, seas and fish farms to detect anomalies or the presence of contaminants. Several parameters can be measured depending on the application:
   – Water quality parameters: pH, oxidation reduction potential, dissolved oxygen, conductivity and temperature.
   – Chemical parameters (ions): ammonium (NH4+), bromide (Br-), calcium (Ca2+), chloride (Cl-), copper (Cu2+), fluoride (F-), iodide (I-), and more.

■ Optimization of electrical and water consumption systems to activate alerts, control consumption and open or close valves in the case of system failures.

■ Monitoring agricultural crops and facilities to reduce water, energy or fertilizer costs, to decrease crop losses and to improve farmers' working conditions. The following parameters can be controlled with the agriculture product line: soil temperature and moisture, leaf wetness, luminosity, and weather values such as temperature, air humidity, pressure, rain and wind.

■ Optimization of industrial processes such as logistical control, automation of manufacturing processes or building management.

■ Monitoring of body parameters easing universal access to a healthcare system for distant communities. Our eHealth platform is equipped with a large number of sensors to measure the most important biometric parameters such as pulse, breath rate, oxygen in blood, temperature, electrocardiogram signals, blood pressure, muscle electromyographic signals, glucose levels, galvanic skin response, lung capacity, snore waves, patient position, airflow and body scale parameters. An alarm button has also been included to connect the user with emergency services.

3. **Do Libelium's sensors ensure security and privacy of contextual data for its users?**

Security and privacy are controversial points in our sector. We really think that should not be considered just as a check point in a conversation with customers. In fact, they just want to know that technology is secure, but they never prove encryption.

The more secure you want your sensor network, the more complex and costly it will be to develop.

Security is a key point for Libelium. For this reason, we added AES 256 libraries for privacy/confidentiality, RSA 1024 for authentication and data integrity, and even an HW encryption circuit in Waspmote to be able to encrypt the information without the need of external SW units.

Security in the IoT is one of the main concerns for municipalities as Smart Cities projects start being a reality. In the following article, published on Libelium website, we expose what is the security proposal developed by Libelium for its sensor devices and gateways with 5 levels of security that includes the Device to Gateway (Symmetric Encryption), Device to Cloud (Symmetric Encryption) and Gateway to Cloud (Secure Web Server – SSL-HTTPS-SSH).

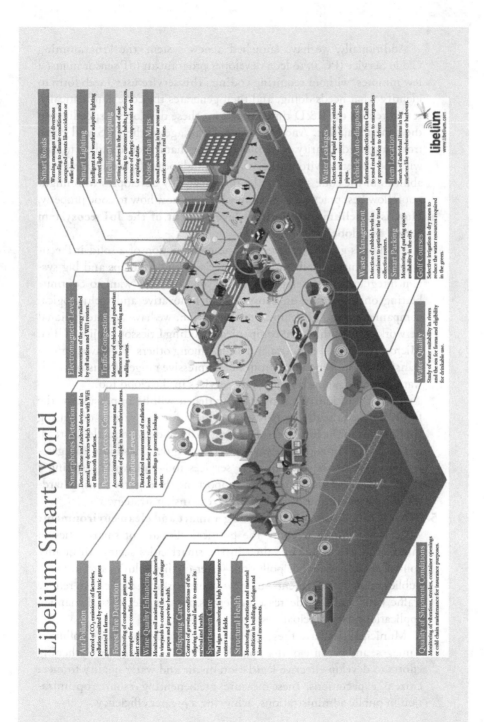

Libelium Smart World

Air Pollution
Control of CO₂ emissions of factories, pollution emitted by cars and toxic gases generated in farms.

Forest Fire Detection
Monitoring of combustion gases and preemptive fire conditions to define alert zones.

Wine Quality Enhancing
Monitoring soil moisture and trunk diameter in vineyards to control the amount of sugar in grapes and grapevine health.

Offspring Care
Control of growing conditions of the offspring in animal farms to ensure its survival and health.

Sportsmen Care
Vital signs monitoring in high performance centers and fields.

Structural Health
Monitoring of vibrations and material conditions in buildings, bridges and historical monuments.

Quality of Shipment Conditions
Monitoring of vibrations, strokes, container openings or cold chain maintenance for insurance purposes.

Smartphones Detection
Detect iPhone and Android devices and in general, any devices which works with WiFi or Bluetooth interfaces.

Perimeter Access Control
Access control to restricted areas and detection of people in non-authorized areas.

Radiation Levels
Distributed measurement of radiation levels in nuclear power stations surroundings to generate leakage alerts.

Electromagnetic Levels
Measurement of the energy radiated by cell stations and WiFi routers.

Traffic Congestion
Monitoring of vehicles and pedestrian affluence to optimize driving and walking routes.

Smart Roads
Warning messages and diversions according to climate conditions and unexpected events like accidents or traffic jams.

Smart Lighting
Intelligent and weather adaptive lighting in street lights.

Intelligent Shopping
Getting advices in the point of sale according to customer habits, preferences, presence of allergic components for them or expiring dates.

Noise Urban Maps
Sound monitoring in bar areas and centric zones in real time.

Water Leakages
Detection of liquid presence outside tanks and pressure variations along pipes.

Vehicle Auto-diagnosis
Information collection from CanBus to send real time alarms to emergencies or provide advice to drivers.

Item Location
Search of individual items in big surfaces like warehouses or harbours.

Waste Management
Detection of rubbish levels in containers to optimize the trash collection routers.

Smart Parking
Monitoring of parking spaces availability in the city.

Golf Courses
Selective irrigation in dry zones to reduce the water resources required in the green.

Water Quality
Study of water suitability in rivers and the sea for fauna and eligibility for drinkable use.

libelium
www.libelium.com

Additionally, we have launched a new system, the Programming Cloud Service (PCS), to let a developer program an IoT sensor in just a few minutes, without requiring coding. This service uses a web form to identify what needs doing, and then generates the code automatically. It will reduce the R&D time needed for these developers, speeding up the time to market. Another major advantage of the PCS will come from improved security as we provide code that is secure and up to date with the current environment, which should mean that security vulnerabilities in the code shouldn't be introduced by developers who might not follow best practices—or simply do not know how to code properly.

4. **How is Libelium guiding the development of the IoT ecosystem across the globe?**

Libelium's growth is based on the construction of a solid IoT ecosystem for mutual collaboration, especially with startups and big system integrators to develop new smart solutions. We want to continue betting on generating an ecosystem of innovative and technological companies around Libelium. For this reason, we have an international network of more than 100 companies including Ericsson, Fujitsu, IBM, Microsoft, NEC, Telefonica and Telit, among others. We have begun a transformation, expanding from our impressive range of sensing hardware and now supporting software services in the cloud too.

It is a transition. We simply know that it will be impossible in the future to provide all the hardware needed for the variety of sensing applications in the IoT, and that trying to do so does not scale. Instead, our focus is to provide gateway services using our Meshlium IoT gateway, and to expand our software services too—all while fostering a cloud partner ecosystem that can support the third-party sensing hardware. It's a significant evolution in the expansion strategy.

5. **What is Libelium's role in creating a smart and clean environment?**

Libelium is now keeping the political climates of smart cities in mind. And this is not in vain—in the smart cities sector, air-quality control sensors to reduce pollution, water-quality solutions to improve public supply management, and new smart-parking solutions to reduce traffic jams and also decrease gas emissions are the most demanded applications by municipalities.

Municipalities have these days a greater awareness about pollution damages, and because of that, public authorities are also strengthening efforts to develop effective legislation on air and water quality to meet citizens' expectations. These measures are benefiting resource optimization in public administrations, achieving a greater efficiency.

As Libelium has gained deeper knowledge on the growing IoT landscape, we have realized that the only pathway is going from lab to reality. In this way, we are putting technology in the hardest scenarios: rain forests and even volcanoes. It is the only one-way road to be aware of trend markets and to align expectations with customers. This is how we have realized that devices need more accuracy, and we have incorporated it without losing sight of the easy installation and maintenance that is also needed to offer the simplest and lowest CAPEX solution.

When the IoT started, the number of nodes was a key point. Having a 1000-node network in a smart city sounded cooler than a 20-node one. Not anymore. Since governments and municipalities got into the game, the accuracy of the information is the new key point. We have seen this tendency clearly in the last few years, since customers have demanded from us new, enhanced products such as calibrated gas sensors, calibrated sound/noise level probes, or even having the possibility to recalibrate the sensor probes themselves (smart water calibration kits and ion sensor calibration kits).

Waggle—An Open Platform for Intelligent Attentive Sensors

Waggle "is a research project at Argonne National Laboratory to design, develop, and deploy a novel wireless sensor platform with advanced edge computing capabilities to enable a new breed of sensor-driven environmental science and smart city research."[*] The project derived its name from "nature's wireless sensors—honeybees. Bees search far and wide for pollen, and report their findings back to the hive using a sophisticated dance called a 'waggle dance'. The dance encodes the distance and angle to the food source."[†]

Waggle has a modular and scalable architecture developed with open-source software and allows adding sensors, computing pipelines and data analytics as needed. The software and hardware designs of the Waggle project are being used by the Array of Things project for building a smart city in Chicago with urban sensors and open data.

The Waggle architecture leverages low-power processors, sensors and cloud computing to build powerful and reliable sensor nodes to actively analyze and respond to data. The key design features of Waggle are security, privacy, extensibility and survivability.

---

[*] Waggle. (n.d.). Retrieved from https://wa8.gl/
[†] Ibid.

Digital Communication Media for IoT

The communication media for 'Things' can be wired or wireless and can be chosen from a variety of protocols. Here we discuss some of the popular communication options.

Internet

The Internet is the most popular digital communication medium. The Internet Protocol (IP) is "a set of technical rules that defines how computers communicate over a network."* Two versions of IP that are currently in use are IP version 4 (IPV4) and IP version 6 (IPV6).

IPV4 was deployed in 1981. It can provide for about four billion Internet addresses. These are 32 bits long. IPV4 addresses consist of a network portion and a host portion that depends on five different address classes: A, B, C, D and E.

IPV6 was deployed in 1999. It can provide for about 2^{128} Internet addresses. IPV6 addresses are 128 bits long, with 64 bits for the network and 64 bits for the host.

Due to global demand for Internet addresses over last two decades, IPV4-based Internet addresses are being consumed at a fast rate. As a result, the inventory of available IPV4 addresses will eventually be exhausted. The projected growth of 'Things' will require more Internet addresses to connect these 'Things' over the Internet. To meet this demand, we have to utilize IPV6-based Internet addresses that allow for a much larger address pool that IPV4. As a result, we will have a mix of IoT devices that are either IPV4-based or IPV6-based, or have an option for either. This might lead to interoperability issues for IoT devices. Standardization in the design of IoT devices can address this issue.

6LoWPAN

The Low-Power Wireless Personal Area Network (LoWPAN) is a low-cost network that "allows wireless connectivity in applications with limited power and relaxed throughput requirements."† 6LoWPAN is an acronym for 'IPV6 over Low-Power Wireless Personal Area Network'. This networking technology allows IPV6 packets to be carried efficiently by devices conforming to IEEE 802.15.4 standards. 6LoWPAN network devices are characterized by their short range, low bit rate and low power consumption. An example of such devices are wireless sensors that can work together to create large mesh networks and connect the physical environment to real-world applications.

---

* ARIN. (n.d.). Retrieved from https://www.arin.net/knowledge/ipv4_ipv6.pdf
† Kushalnagar, N., Montenegro, G., & Schumacher, C. (2007). IPv6 over low-power wireless personal area networks (6lowpans): Overview, assumptions, problem statement, and goals. Retrieved from https://tools.ietf.org/html/rfc4919

Zigbee

Zigbee* is a standard for low-power mesh networks based on IEEE 802.15.4 standards. It can be used in indoor as well as outdoor IoT solutions. The first Zigbee specification was made available by the Zigbee Alliance in 2005. Zigbee 3.0 allows wireless interoperability of products from different manufacturers who are approved through a certification scheme. Zigbee operates in the 2.4 GHz radio band, which is available for use globally without a license, so applications using Zigbee are portable to any global location. Non-routing devices using the Zigbee standard can run on power supplies like batteries or solar cells, or can utilize Zigbee Green Power. For secured, over-the-air transfer of information, Zigbee utilizes AES128 encryption.

Bluetooth Low Energy

Bluetooth Low Energy† (BLE) is a wireless personal area network designed by the Bluetooth Special Interest Group. It can enable short-burst wireless connections in various network topologies, like the point-to-point (P2P) topology for one-to-one device communication, the broadcast topology for one-to-many device communication or the mesh topology for many to many device communications. BLE supports major mobile-computing platforms like iOS, Android, Windows, Linux. The BLE broadcast topology can be utilized for localized information sharing, such as item-finding beacons in smart retail solutions. BLE P2P is ideal for connected devices like fitness trackers and health monitors. The BLE mesh can be used in smart solutions like asset tracking and building automation.

LoRaWAN

It is a Low Power Wide Area Network (LPWAN) for battery operated wireless devices with features like bi-directional communication, localization services and mobility. It can be used in regional, national or global networks. Long Range Low Power Wide Area Network (LoRaWAN)‡ uses gateways to relay messages between end devices and a central network server. On LoRaWAN, the end devices use single-hop wireless communication, while gateways connect using standard IP connections. LoRaWAn utilizes an unlicensed radio spectrum for communication and AES128 encryption for the security of transmitted data. This network can be used in smart city applications such as low-power tracking applications that are GPS-free and cloud-based data delivery to mobile devices and smart systems.

---

* Zigbee Alliance. (n.d.). Retrieved from http://www.zigbee.org/what-is-zigbee/
† Specifications. (n.d.). Retrieved from https://www.bluetooth.com/specifications
‡ LoRa Alliance™. (n.d.). Retrieved from https://lora-alliance.org/technology

Modbus

Modbus is a serial communication protocol* for industrial devices and an enabler for the Industrial IoT (IIoT). It is an open protocol and follows a master-slave model whereby the 'master' device requests information and the 'slave' device supplies the information. Modbus can be used in supervisory control and data acquisition systems.

MQTT

Message Queuing Telemetry Transport (MQTT)[†] is an open, machine-to-machine connectivity protocol for IoT communication. It is a lightweight protocol, having MQTT broker as mediator for interacting MQTT agents. MQTT follows a publication-subscription model whereby the MQTT agents publish information that are consumed by the subscribers. This is implemented through the MQTT methods—connect, disconnect, publish, subscribe and unsubscribe.

Cloud, Fog and Data Analytics

As sensors collect contextual data, based on their design and architecture, this data can be processed locally on a smart device to a certain extent or can be flushed intermittently to a gateway device for zonal processing. This computing at the edge of a network is termed as 'edge'[‡] or 'fog'[§] computing. The data from sensors can also be sent to a cloud-based storage and processing location. Cloud services can be public, private, or a communal hybrid or the two or community, based on the smart service's architecture and design.

Depending on the need of the smart service rendered, the IoT application may be integrated with a data analytics engine for fine tuning and customization of the application output.

For example, domain-based IoT applications can be enabled with various infrastructure and functional components, such as sensors that capture contextual data based on predefined parameters, gateway devices that gather data from a bunch of sensors, data storage that can be at the edge or hosted in the cloud where the gateway devices flush the gathered data intermittently, analytical processing functions, application programming interface (API)-based business functions, command and control functions for the actuators in sensors, and wired or wireless network communications connecting these components, as shown in Figures 2.1 and 2.2.

---

* J., M. T. (2016, March 31). A comparison of IoT gateway protocols: MQTT and modbus. Retrieved from https://software.intel.com/en-us/articles/a-comparison-of-iot-gateway-protocols-mqtt-and-modbus

† MQTT. (n.d.). Retrieved from http://mqtt.org/

‡ Open Edge Computing. (n.d.). Retrieved from http://openedgecomputing.org/

§ OpenFog Consortium (n.d.). Retrieved from https://www.openfogconsortium.org/

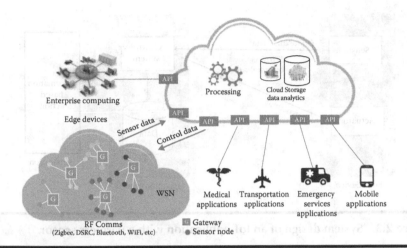

Figure 2.1 Infrastructure and functional components for the Internet of Things.

Cloud Security Alliance. (2015). Security guidance for early adopters of the internet of things. Retrieved from https://downloads.cloudsecurityalliance.org/white papers/Security_Guidance_for_Early_Adopters_of_the_Internet_of_Things.pdf

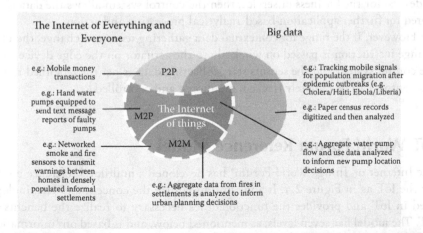

Figure 2.2 Big Data generated from the interaction of IoT, machines and persons.

Biggs, P., Garrity, J., LaSalle, C., & Polomska, A. (n.d.). Harnessing the internet of things for global development. Retrieved from https://www.sbs.ox.ac.uk /cybersecurity-capacity/system/files/Harnessing-IoT-Global-Development.pdf

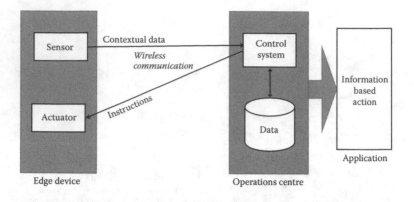

Figure 2.3 System design of an IoT application with a wireless sensor.

Chaudhuri, A. (2016). Cyber threat mitigation of wireless sensor nodes for secured, trustworthy IoT services. The EDP Audit, Control, and Security Newsletter, 54(1), 1–14. doi:10.1080/07366981.2016.1181416

IoT System Design

If we look at the system design of an IoT application with wireless sensor nodes, as in Figure 2.3, we will see that the data collected by the sensor is sent wirelessly to an operations center that has a control system to monitor the relevance of collected data as per the contextual requirement. If the data are within the required range as desired for the business or service, then the control system allows the data to be stored for further application-based analytical processing and action.

However, if the range for contextual data gathering requires a change, then the change instruction is passed on wirelessly to the actuator on the edge device from the control system and the sensors start collecting data as per the redefined range. The sensors can be remotely tracked, monitored and controlled.

IoT World Forum Reference Model

The Internet of Things World Forum* has developed a multilevel reference model for the IoT as in Figure 2.4. It aims to standardize the concept and terminology used in IoT and provides the functionalities necessary to realize the benefits of IoT. The model has seven levels, as mentioned below, and is based on 'information flow'.

---

* Internet of Things World Forum. (n.d.). Retrieved from https://www.iotwf.com/

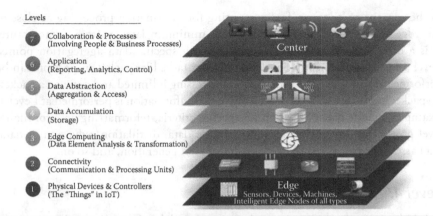

Figure 2.4   IoT World Forum Reference Model CISCO.

CISCO (2014). The internet of things reference model. Retrieved from http://cdn .iotwf.com/resources/71/IoT_Reference_Model_White_Paper_June_4_2014.pdf

Level 1: Physical Devices and Controllers

Physical devices and controllers are the endpoint devices that send and receive information, and are the 'Things' in the IoT. The devices can be queried or controlled over the Internet and are capable of analog-to-digital conversion of signals as needed and contextual data generation. Due to low computing and storage power, the devices will flush out captured data intermittently in small units to the networking equipment in Level 2. The controller will control the data parameters based on authorized instructions.

Level 2: Connectivity

The communication and connectivity of the 'Things' is maintained at this level for reliable and timely information transmission. Information can flow between reliable devices in Level 1 and the network, across networks (east-west traffic), and between networks (Level 2) and low-level information processing that occurs at Level 3. Communication gateways can be introduced to connect legacy devices that are not IP enabled. Some computation activities like applying network security policies or protocol translation can occur at Level 2.

Level 3: Edge (Fog) Computing

There might be requirements for localized conversion of network data flows into information to cater to specific IoT service needs. In such situations, it can be an operational overhead as well as being time-consuming to send contextual data from sensor devices to a centralized 'cloud' for processing and further action. This leads to the concept of 'edge' or 'fog' computing for information processing as close to the edge of the network as possible, with minimum latency from data capture. It is a decentralized computing approach with specific data aggregation points. Level 3 focus on such activities. Computation tasks like packet inspection can be performed at this level. The information processing is limited and done on a packet-by-packet basis. Higher level processing of this information is performed at Level 4. Examples include data evaluation for specific criteria, reformatting data for higher level processing, expanding/decoding cryptic data, distillation/reduction of data, data assessment for threshold attainment or alert generation, and so on.

Level 4: Data Accumulation

This is the storage level where in-motion, event-based data from a network is converted to data at rest for query-based processing by applications when necessary, on a non-real-time basis. Some of the activities performed at this level include event filtering/sampling, event comparison, event aggregation and northbound/southbound alerting.

Level 5: Data Abstraction

This level helps in data aggregation from multiple devices and simplifies access of data to the application by creating schemas and views of data. The key processes at this level are filtering, selecting, projecting and reconciliation of data in different formats, semantics consistency of data from different sources, normalizing/de-normalizing and indexing of data.

Level 6: Application

All kinds of applications reside at this level, which provides the designed output by interpretation of available information. These may be critical business applications, mobile applications, business intelligence reports, analytics, control applications and so on.

Level 7: Collaboration and Processes

This level deals with the people and business processes for communication and collaboration that are necessary to make the IoT application useful.

The Industrial Internet of Things and System Architecture

The 'Industrial IoT' (IIoT) refers to the convergence of the industrial ecosystem, contextual sensing, computing and ubiquitous network connectivity. The Industrial

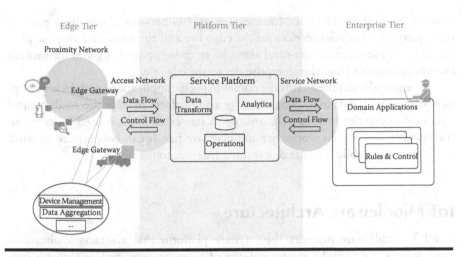

Figure 2.5   Three-tiered Industrial IoT Systems Architecture.

Industrial Internet Consortium. (2015). Industrial internet reference architecture. Retrieved from https://www.iiconsortium.org/IIRA-1-7-ajs.pdf

Internet Consortium (IIC) refers to the IIoT as "the Internet of things, machines, computers and people, enabling intelligent industrial operations using advanced data analytics for transformational business outcomes."[*]

The IIoT creates Industrial Internet Systems (IISs) by connecting the industrial control systems (ICS) online with enterprise systems, business processes, analytics solutions and humans. Examples include industrial systems for healthcare, energy, the public sector, transportation, manufacturing and so on. Safety, security and resilience are the primary characteristics of the IIoT and IISs.

The IIC has developed a system architecture for IIoT applications.[†] It is a three-tiered architecture as shown in Figure 2.5, with edge, platform and enterprise tiers that are connected by three networks—a proximity network, an access network and a service network respectively.

The edge comprises the edge nodes. Data is collected from the edge nodes and communicated using the proximity network. The proximity network connects the edge nodes to an edge gateway that connects to other networks. Depending on the storage and computation capacity, some data aggregation, processing and analytics may be performed at the edge gateway, and it can be used as a management point for the devices and assets.

The platform tier is the middle tier of the three-tiered architecture, and uses the access network and service network to communicate with the edge tier and enterprise tier respectively. The access network can be a corporate network or a private

---

[*] Industrial Internet Consortium. (2015). Industrial internet reference architecture. Retrieved from https://www.iiconsortium.org/IIRA-1-7-ajs.pdf
[†] Ibid.

network overlaid on the public Internet or a 4G/5G network. Apart from proving data-query and analytics services for the edge tier and the enterprise tier, the platform tier also manages devices and assets by receiving, processing and forwarding control commands from the enterprise tier to the edge tier.

The enterprise tier maintains the domain-specific applications and decision support systems with data input from the edge and platform tiers. It also sends control data to the edge tier and the platform tier. A service network is used for communication between the enterprise tier and platform tier. It can be a private network overlaid on the public Internet or a secured Internet connection.

IoT Middleware Architecture

The IoT middleware provides the software platform that abstracts applications from devices and provides interoperability of heterogeneous devices through syntactic and semantic associations. This functional layer is also responsible for device authentication, security and privacy of the contextual data and data collection and exchange in volumes with applications to render the designed IoT service.

We discuss here the LinkSmart* middleware architecture for IoT to provide an insight about the necessary components for building a robust middleware. It is an output from two integrated European research projects: Hydra† and EBBITS. LinkSmart is an open-source middleware platform for networked-embedded systems and IoT applications.

The LinkSmart middleware constitutes a software layer between the operating system of software-enabled devices and user applications that communicate with those devices. It provides protocols that execute on top of the transport layer and provide services to the application layer.

The nine technical components in the LinkSmart architecture are:

1. Service-oriented architecture
2. Model-driven approach
3. Three-layered discovery architecture
4. P2P-based network architecture
5. Dynamic runtime architecture
6. Context management
7. Self-management features comprising of goal management, change management and component control
8. Security and trust
9. Storage management

---

* LinkSmart®. (n.d.). Retrieved from https://docs.linksmart.eu/display/HOME/What+is+Link Smart

† Hydra. (n.d.). Retrieved from http://www.hydramiddleware.eu/articles.php?article_id=68

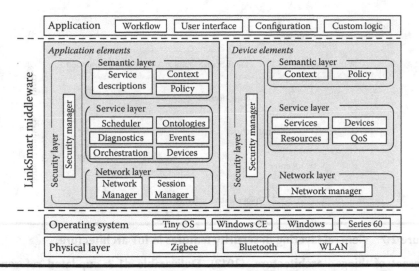

Figure 2.6   Structural overview of the LinkSmart middleware layers.

Kostelnik, P., Sarnovsky, M., & Furdík, K. (2011). The semantic middleware for networked embedded systems applied in the internet of things and services domain. Scalable Computing: Practice and Experience, 12(3), 307–315. Retrieved from http://ebbits-project.eu/downloads/papers/semantic_middleware.pdf

The middleware, devices and services are integrated in a service-oriented architecture, which effectively turns all devices into web services and provides extensive syntactic interoperability so that the components can talk to each other regardless of their physical locations and the interface technology.

As depicted in Figure 2.6, the LinkSmart middleware architecture has four layers—a semantic layer, a service layer, a network layer and a security layer—which are designed separately for application elements and devices.

The semantic layer provides service descriptions, context and policy. The service layer is responsible for the scheduling of jobs, diagnostics and the orchestration of application elements and the resource optimization of device elements for a seamless integration of services. The network layer manages the network and sessions, while the security of application elements and devices is managed through the security layer.

The modular architecture of LinkSmart provides the flexibility to create any network of devices necessary to build an IoT application.

IoT Security Architecture

Security is a prime necessity for IoT devices and services. All components in IoT architecture should be secured to provide trustworthy services. To achieve

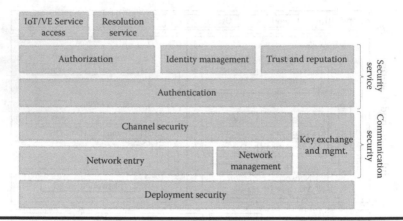

Figure 2.7 General layering of security features in IoT architecture.

Internet of things—architecture. (2012). Deliverable d1.3—updated reference model for IoT v. 1.5. Retrieved from http://cocoa.ethz.ch/downloads/2014/01 /1524_D1.3_Architectural_Reference_Model_update.pdf

this, IoT-A,* the European Lighthouse Integrated Project on IoT architecture, has recommended a layered security approach for IoT architecture, as depicted in Figure 2.7. As per IoT-A terminology, physical entities are represented in the digital world as virtual entities (VE).

The security architecture is layered into three key areas: deployment security, communication security and service security. Devices (sensor, actuator), resources (network resource, on-device resource) and services should be deployed considering all kinds of threat scenarios and security needs. The communication security layer considers all aspects of threats in communication among devices, resources and services, with emphasis on communication channel security, network security and management and key exchange and management. Gateways can play a critical role in secured communication between an unconstrained device network and a constrained device network through communication-protocol adaptation and security-configuration management in peripheral networks of constrained devices. The service security layer ensures authentication, authorization and identity management for virtual entities, including auto-ID devices, and for secured access to resolution-service components.

---

* European Commission. (n.d.) CORDIS. Retrieved from https://cordis.europa.eu/project /rcn/95713_en.html

Cyber-Physical Systems and IoT

According to NIST:

> Cyber-physical systems (CPS) are smart systems that include engineered interacting networks of physical and computational components. These highly interconnected and integrated systems provide new functionalities to improve quality of life and enable technological advances in critical areas, such as personalized health care, emergency response, traffic flow management, smart manufacturing, defense and homeland security, and energy supply and use. In addition to CPS, there are many words and phrases (Industrial Internet, Internet of Things (IoT), machine-to-machine (M2M), smart cities, and others) that describe similar or related systems and concepts. There is significant overlap between these concepts, in particular CPS and IoT, such that CPS and IoT are sometimes used interchangeably.[*]

According to NIST:

> A CPS generally involves sensing, computation and actuation. CPS involve traditional information technology (IT) as in the passage of data from sensors to the processing of those data in computation. CPS also involve traditional operational technology (OT) for control aspects and actuation. The combination of these IT and OT worlds along with associated timing constraints is a particularly new feature of CPS.[†]

However, IoT and CPS have some conceptual distinction according to some. One such distinction mentions[‡] CPS as a tight human-machine interaction that provides control of combined organizational and physical processes. An example of this is a networked, distributed traffic-management system. On the other hand, IoT is an enabler for sensing the physical world through Internet connectivity, like in a smart-city transportation system.

---

[*] NIST. (2016). Framework for cyber-physical systems; release 1.0. Retrieved from https://s3.amazonaws.com/nist-sgcps/cpspwg/files/pwgglobal/CPS_PWG_Framework_for_Cyber_Physical_Systems_Release_1_0Final.pdf

[†] Ibid.

[‡] Soldatos, J. (2015, December 30). IoT vs. m2m, cps, wot....: are these terms synonyms? Retrieved from https://www.linkedin.com/pulse/iot-vs-m2m-cps-wot-terms-synonyms-john-soldatos

Suggested Reading

Atzori, L., Iera, A., Morabito, G., & Nitti, M. (2012). The social internet of things (siot)—when social networks meet the internet of things: Concept, architecture and network characterization. *Computer Networks, 56*(16), 3594–3608.

Carrez, F., Elsaleh, T., Gómez, D., Sánchez, L., Lanza, J., & Grace, P. (2017, June). A reference architecture for federating IoT infrastructures supporting semantic interoperability. In *2017 European Conference on Networks and Communications (EuCNC)* (pp. 1–6). IEEE.

Catarinucci, L., De Donno, D., Mainetti, L., Palano, L., Patrono, L., Stefanizzi, M. L., & Tarricone, L. (2015). An IoT-aware architecture for smart healthcare systems. *IEEE Internet of Things Journal, 2*(6), 515–526.

Chakrabarty, S., & Engels, D. W. (2016, January). A secure IoT architecture for smart cities. In *Consumer Communications & Networking Conference (CCNC), 2016 13th IEEE Annual* (pp. 812–813). IEEE.

Cirani, S., Picone, M., Gonizzi, P., Veltri, L., & Ferrari, G. (2015). IoT-oas: An oauth-based authorization service architecture for secure services in IoT scenarios. *IEEE Sensors Journal, 15*(2), 1224–1234.

Datta, S. K., Bonnet, C., & Nikaein, N. (2014, March). An IoT gateway centric architecture to provide novel M2M services. In *Internet of Things (WF-IoT), 2014 IEEE World Forum* (pp. 514–519). IEEE.

Espinoza, J. R., Padilla, V. S., & Velasquez, W. (2017). IoT generic architecture proposal applied to emergency cases for implanted wireless medical devices. In *Proceedings of the International Multi-Conference of Engineers and Computer Scientists* (Vol. 2).

Fortino, G., Guerrieri, A., Russo, W., & Savaglio, C. (2014, May). Integration of agent-based and cloud computing for the smart objects-oriented IoT. In *Computer Supported Cooperative Work in Design (CSCWD),* Paper presented at Proceedings of the 2014 IEEE 18th International Conference (pp. 493–498). IEEE.

Furdik, K., Lukac, G., Sabol, T., & Kostelnik, P. (2013). The network architecture designed for an adaptable IoT-based smart office solution. *International Journal of Computer Networks and Communications Security, 1*(6), 216–224.

Ganchev, I., Ji, Z., & Odroma, M. (2014). A generic IoT architecture for smart cities. *25th IET Irish Signals & Systems Conference 2014* and *2014 China-Ireland International Conference on Information and Communities Technologies (ISSC 2014/CIICT 2014).* doi:10.1049/cp.2014.0684

Google Cloud. (n.d.). Retrieved from https://cloud.google.com/solutions/architecture/real-time--stream-processing-iot

Granados, J., Rahmani, A. M., Nikander, P., Liljeberg, P., & Tenhunen, H. (2014, November). Towards energy-efficient healthcare: An internet-of-things architecture using intelligent gateways. In *2014 EAI 4th International Conference on Wireless Mobile Communication and Healthcare (Mobihealth)* (pp. 279–282). IEEE.

Intel. (n.d.). Reference architecture: Securing edge to IoT cloud solutions with Intel and Amazon web services. Retrieved from https://www.intel.com/content/dam/www/public/us/en/documents/reference-architectures/securing-edge-cloud-iot-solutions-aws-architecture.pdf

Intel. (n.d.). IoT platform reference architecture. Retrieved from https://www.intel.com/content/www/us/en/internet-of-things/white-papers/iot-platform--reference-architecture-paper.html

Internet of Things Expert Group. (2013). IoT architecture. Retrieved from http://ec.europa
.eu/information_society/newsroom/cf/dae/document.cfm?doc_id=1750

Istepanian, R. S., Hu, S., Philip, N. Y., & Sungoor, A. (2011, August). The potential of internet of m-health Things "m-IoT" for non-invasive glucose level sensing. In *Engineering
in Medicine and Biology Society*. Paper presented at the 2011 Annual International
Conference of the IEEE (pp. 5264–5266). IEEE.

Kantarci, B., & Mouftah, H. T. (2014). Trustworthy sensing for public safety in cloud-
centric internet of things. *IEEE Internet of Things Journal*, *1*(4), 360–368.

Kovatsch, M., Mayer, S., & Ostermaier, B. (2012, July). Moving application logic from the
firmware to the cloud: Towards the thin server architecture for the internet of things.
In *Innovative Mobile and Internet Services in Ubiquitous Computing*. Paper presented
at the 2012 Sixth International Conference (pp. 751–756). IEEE.

Lloret, J., Tomas, J., Canovas, A., & Parra, L. (2016). An integrated IoT architecture for
smart metering. *IEEE Communications Magazine*, *54*(12), 50–57.

Ma, J., Nguyen, H., Mirza, F., & Neuland, O. (2017). Two way architecture between IoT
sensors and cloud computing for remote health care monitoring applications. In
Proceedings of the 25th European Conference on Information Systems (pp. 2834–2841).

Microsoft. (2017 Oct. 11). Azure and the Internet of Things. Retrieved from https://docs.micro
soft.com/en-us/azure/iot-suite/iot-suite-what-is-azure-iot#iot-solution-architecture

Ning, H., & Liu, H. (2012). Cyber-physical-social based security architecture for future
internet of things. *Advances in Internet of Things*, *2*(01), 1.

Nitti, M., Pilloni, V., Giusto, D., & Popescu, V. (2017). IoT architecture for a sustainable tour-
ism application in a smart city environment. *Mobile Information Systems*, *2017*, 1–10.

Perles, A., Pérez-Marín, E., Mercado, R., Segrelles, J. D., Blanquer, I., Zarzo, M., & Garcia-
Diego, F. J. (2017). An energy-efficient internet of things (IoT) architecture for preventive conservation of cultural heritage. *Future Generation Computer Systems*, *81*, 566–581.

Rhee, S., Nguyen, C. T., Nelson, A., Hoffman, D., & Toth, G. (2017, April). Towards
a foundation for a collaborative replicable smart cities IoT architecture. In *Second
International Workshop on Science of Smart City Operations and Platforms Engineering
(SCOPE)*.

Schmid, S., Bröring, A., Kramer, D., Käbisch, S., Zappa, A., Lorenz, M., . . . & Gioppo, L.
(2016, November). An architecture for interoperable IoT ecosystems. In *International
Workshop on Interoperability and Open-Source Solutions*, (pp. 39–55). Springer, Cham.

Sowe, S. K., Kimata, T., Dong, M., & Zettsu, K. (2014, July). Managing heterogeneous
sensor data on a big data platform: IoT services for data-intensive science. In *Computer
Software and Applications Conference Workshops (COMPSACW)*. Paper presented at
the 2014 IEEE 38th International (pp. 295–300). IEEE.

Ungurean, I., Gaitan, N. C., & Gaitan, V. G. (2014, May). An IoT architecture for things
from industrial environment. In *Communications (COMM)*. Paper presented at the
2014 10th International Conference on (pp. 1-4). IEEE.

Weyrich, M., & Ebert, C. (2016). Reference architectures for the internet of things. *IEEE
Software*, *33*(1), 112–116.

Yashiro, T., Kobayashi, S., Koshizuka, N., & Sakamura, K. (2013, August). An internet
of things (IoT) architecture for embedded appliances. In *Humanitarian Technology
Conference (R10-HTC), 2013 IEEE Region 10* (pp. 314–319). IEEE.

Zhang, H., & Zhu, L. (2011, June). Internet of things: Key technology, architecture and
challenging problems. In *International Conference Computer Science and Automation
Engineering (CSAE), 2011 IEEE* (Vol. 4, pp. 507–512). IEEE.

Chapter 3

The Philosophy of Information and Ethics in the Internet of Things Technology

Information is a distinction that makes a difference.

Donald M. MacKay

After reading this chapter you will be able to:

- Grasp the basics of philosophy of information in relation to IoT
- Understand the techno-philosophical aspects of IoT technology
- Gain an insight on digital ethics and algorithmic accountability
- Understand the ethical concerns of IoT applications
- Interpret the IoT belief system
- Understand the relation of attention, subjectivity, objectivity and happiness with IoT technology

Introduction

The philosophy of information has been defined by eminent philosopher and professor L. Floridi as "the philosophical field concerned with a) the critical investigation of the conceptual nature and basic principles of information, including its dynamics, utilization, and sciences, and b) the elaboration and application of information-theoretical and computational methodologies to philosophical problems."*

In the context of IoT technology, the process of information generation, use and its afferent and efferent flows in various smart devices, applications and services, are critical for the creation of a vibrant and dynamic IoT ecosystem. If we can understand the life-cycle of information in an IoT landscape and use that information with due consideration of various factors like ownership, consent and regulatory compliance, then we can create trust in this emerging technology. The trust will enhance user confidence and reputation, resulting in greater adoption of this potential technology for personal and social benefits.

This chapter discusses the techo-philosophical aspects of IoT technology, covering the concepts of attention, subjectivity, objectivity and happiness. It raises key ethical concerns and addresses the needs for algorithmic transparency and accountability in autonomous IoT applications and an ethical IoT ecosystem to build an IoT belief system for mass acceptance of this potential technology. The interplay of smart devices, smart services and humans has been showcased here to highlight the trust permeability in human and IoT interfaces.

The Philosophical Dimensions of IoT Technology

An information society builds on the belief system of its people who create, use and cater to the information needs by immersing themselves in the 'cyber infosphere' through the process of adoption, usage, and knowledge exchange with other users resulting in digital evolution of self as well as the information systems, in the process.

To build trust in IoT technology in an information society, we have to provide satisfactory answers to techno-philosophical queries from users, which can be broadly categorized under five dimensions as discussed below and depicted in Figure 3.1.

Dimension 1: Ontology

According to T.R. Gruber, Ontology is "an explicit specification of a conceptualization" and "a systematic account of Existence. For knowledge-based systems, what "exists" is exactly that which can be represented."† For IoT technology, ontology specifies the existence of the IoT in the form of smart devices, network communication, application-based contextual data capture, data flow between systems and

---

* Floridi, L. (2011). The philosophy of information. New York: Oxford University Press.
† Gruber, T.R. (1993). A translation approach to portable ontology specifications. *Knowledge Acquisition*, 5(2), 199–220. doi:10.1006/knac.1993.1008

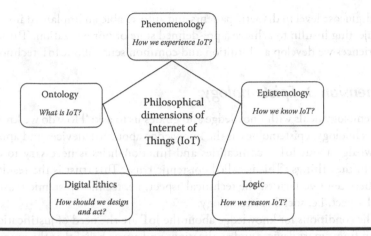

Figure 3.1   The five dimensions of Philosophy of the IoT technology.

conditioned action. With the IoT being an emerging technology, its first-time users need a clear interpretation of it as an initial confidence-building step toward further usage and adoption. In an ontological sense, we have to ask the classical question by Socrates: 'Ti esti?' For the IoT, that is: 'What is the IoT?'

Dimension 2: Phenomenology

Once the potential end user interprets the existence of IoT technology, the next step is to ask: 'How do we experience it?' Various IoT applications, like smart-room ambience devices that automatically gather contextual data and change the ambience in a room through pre-set algorithmic rules, smart monitoring devices that alert the nearest police station about security violation in a locality, and others build consciousness about IoT technology among users through context-based conditioned response and action mechanism. According to Professor T.D. Wilson, the aim of phenomenology is "to study how human phenomena are experienced in consciousness, in cognitive and perceptual acts, as well as how they may be valued or appreciated aesthetically. Phenomenology seeks to understand how persons construct meaning."*

Smart sensors help to build the phenomena of autonomous perception, intentionality and embodiment when we apply IoT technology for awareness-based actions, like auto-enabling water sprinklers on plants in a garden based on relative humidity in the air and the dryness of leaves, or when smart sensors perceive the

---

* Wilson, T. D. (2002). Alfred Schutz, phenomenology and research methodology for information behaviour research. *The New Review of Information Behaviour Research, 3*(71), 1–15. Retrieved from http://web.pdx.edu/~tothm/theory/Schutz%20%26%20Research%20 Methodology.docx

blood-glucose level in diabetic patients and auto-enable an implanted insulin pump for injecting insulin to achieve a pre-defined state of normalization. Through these experiences we develop an intuition and common sense about IoT technology.

Dimension 3: Epistemology

Epistemology deals with knowledge and guides us to ask: 'How do we know it?' For IoT technology, epistemology is the knowledge about IoT devices and applications. Knowledge about IoT technicalities and functionalities is necessary to develop a trust in the 'Things'. This is called epistemic trust. This trust is the result of interplay between two factors: the technical aspect, i.e. techno-epistemic trust, and the social aspect, i.e. social epistemology.

The conditions for knowledge about the IoT are the need of justification for the output from smart devices and applications and a true belief that the output is what it should be. This implies that to develop epistemic trust in IoT technology, end users should develop a true belief bounded by rationality for proper justification of the techniques and output of the 'Things'.

IoT technology can potentially become socially adopted if humans accept it in various spheres of life. This is a gradual process, as the techno-epistemic knowledge of IoT technology and its benefits of use emerge from the usage of devices and applications, interaction with devices, knowledge exchange with other users, manufacturers and other stakeholders, and adoption. In this process, the social dimension of IoT knowledge is developed. This is called social epistemology. The agents for knowledge creation and dissemination for IoT technology can be humans as well as computing devices (non-humans). Through the process of knowledge development and distribution about IoT technology by human and non-human epistemic agents, a social trust will develop in this emerging technology.

Dimension 4: Logic

With respect to the technological aspects and output of IoT devices and applications, the philosophical dimension of logic helps us ask: 'How do we reason with it?' Logical analysis provides the reasons, and is closely aligned with epistemology, as reasoning is epistemic.

An IoT user will experience the contextual output from the smart device or service and will try to deduce the technical reason behind the output. The basic queries of the user will be related to technical know-how, reliability, security and privacy risks. Deductive reasoning for the user happens through various explicit means, like accessing knowledge references provided by the product or service vendor, access to user forums for knowledge exchange, and other means like self-help portals, regulatory guidance and so on.

Dimension 5: Digital Ethics

Ethics "involves systematizing, defending, and recommending concepts of right and wrong behavior."* For IoT technology, ethics require us to ask: 'How should we design and interact with it?' IoT being a digital technology, we have to consider the ethical aspects in its design and use from a digital perspective. The digital actors are all stakeholders, including those designing IoT devices and services, standardization and governing bodies, vendors as well as users.

We have to rely on normative digital ethics to define moral standards for IoT design and operation by analyzing the impact of reliability, security, data ownership, data traceability, user consent, compliance to legal and regulatory requirements and social values.

Algorithmic Accountability

Accountability in the design of algorithms will help to "ensure that automated decision-making systems remain accountable and comprehensible to the individuals affected by their decisions."† Algorithmic accountability is also a key consideration for ethical use of the IoT in order to prevent biased decision making and erroneous outputs that can be designed to be opaque to users.

The Principles for Algorithmic Transparency and Accountability proposed by the U.S. Association for Computing Machinery (USACM) Public Policy Council‡ are highly relevant for IoT technology and have been provided in Table 3.1 for ready reference.

IoT Belief System

Today we are living in a digital society. The anthropological landscape of human society across the globe is gradually enwrapping itself in a web of information that has been made possible with digital technologies. IoT technology will be a catalyst in digitizing various aspects of human life through potential smart applications like the smart home, smart healthcare, smart transport management and other digital offerings that are being conceived globally. It will create a boundary-less space of things with information flowing among the dependent and interdependent cyber systems.

---

* Internet Encyclopaedia of Philosophy. (n.d.). Retrieved from http://www.iep.utm.edu/ethics/
† Digital Ethics Lab, University of Oxford. (n.d.). Retrieved from http://digitalethicslab.oii.ox
  .ac.uk/ethical-auditing-for-accountable-automated-decision-making/
‡ USACM. (2017). Principles for algorithmic transparency and accountability. Retrieved from
  https://www.acm.org/binaries/content/assets/public-policy/2017_usacm_statement_algo
  rithms.pdf

Table 3.1 USACM Principles for Algorithmic Transparency and Accountability

Principles for Algorithmic Transparency and Accountability
1. **Awareness:** Owners, designers, builders, users, and other stakeholders of analytic systems should be aware of the possible biases involved in their design, implementation, and use and the potential harm that biases can cause to individuals and society.
2. **Access and redress:** Regulators should encourage the adoption of mechanisms that enable questioning and redress for individuals and groups that are adversely affected by algorithmically informed decisions.
3. **Accountability:** Institutions should be held responsible for decisions made by the algorithms that they use, even if it is not feasible to explain in detail how the algorithms produce their results.
4. **Explanation:** Systems and institutions that use algorithmic decision-making are encouraged to produce explanations regarding both the procedures followed by the algorithm and the specific decisions that are made. This is particularly important in public policy contexts.
5. **Data Provenance:** A description of the way in which the training data was collected should be maintained by the builders of the algorithms, accompanied by an exploration of the potential biases induced by the human or algorithmic data-gathering process. Public scrutiny of the data provides maximum opportunity for corrections. However, concerns over privacy, protecting trade secrets, or revelation of analytics that might allow malicious actors to game the system can justify restricting access to qualified and authorized individuals.
6. **Auditability:** Models, algorithms, data, and decisions should be recorded so that they can be audited in cases where harm is suspected.
7. **Validation and Testing:** Institutions should use rigorous methods to validate their models and document those methods and results. In particular, they should routinely perform tests to assess and determine whether the model generates discriminatory harm. Institutions are encouraged to make the results of such tests public.

Privacy and security are key components of digital trust for emerging IoT technology. For social adoption of this technology, we have to build a reputation for IoT offerings. Reputations help to build confidence over a period of time through a belief system. For IoT technology to flourish, we have to build user confidence through a belief system, as depicted in Figure 3.2.

Because of the benefits of smart products and services, users develop reliance on them to meet their specific needs and gradually there is an enhanced confidence of

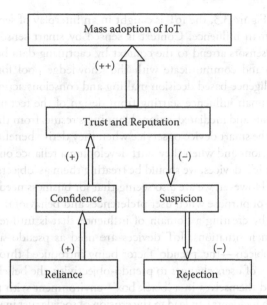

Figure 3.2   The IoT belief system.

association with the smart product or service. This builds trust and improves the reputation of smart services and leads to mass adoption. However, if users have concerns about a smart device or service, and if those concerns are not addressed by product vendor or service provider, then this can lead to suspicion among the users. This can degrade user confidence and ultimately lead to a rejection of the smart device or service.

IoT—Attention, Subjectivity, Objectivity and Happiness

As users experience the various outputs of IoT applications, they develop an ontological and epistemic sense of it. This helps users to build a subjective consciousness about the technology.

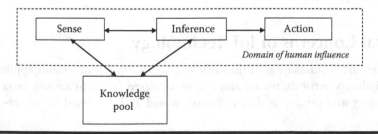

Figure 3.3   Interplay of sense, inference and action under human influence in the cyber infosphere.

As shown in Figure 3.3, the IoT is caught in an interplay of sense, inference and action under human influence. Context is sensed by smart sensors as per defined boundaries. The sensors attend to the context by capturing data bounded by upper and lower limits and communicate with the knowledge pool for inference-based action. This intelligence-based decision making and conscious action happens under the domain of human influence starting from design of the technical components, process components and mediated action. Happiness results from the users' increased association with the smart device or service when they derive benefits from it without concerns or suspicions and when they start developing a reliance on it.

When we use IoT devices, we should be treating them as 'objects of purpose', and users as subjects. However, we are also seeing that for business needs, users are being treated as objects of purpose when user preferences and behavioral data are captured without consent by creating a domain of influence that is unintended from user's perspective. In such situations, IoT devices are used as pseudo-subjects and users become pseudo-objects—the 'pseudo' factor being introduced through obfuscation. As the subjectivity of users changes to pseudo-objectivity, the belief system collapses and the users find themselves in a 'black-box'* environment with these IoT devices and services. Such incidents can lead to the erosion of social trust in smart devices and services and prevent us from realizing the social benefits of this emerging technology.

The interwoven physical and virtual reality of the IoT should be designed in such a manner that humans are not completely removed from the decision-making process for intelligence-based action. We have to ensure the use of appropriate algorithms in IoT design that allow humans to make moral judgments to prevent inference and data privacy breaches, and for informed personal data usage with consent. IoT artifacts can be considered as moral agents or a source of moral action because they can cause moral good or harm to their users and society.

In an IoT-enabled digital society, we have to establish a techno-trust relationship between technical artifacts and their users. We also need a reputation system for the IoT to provide explicit qualitative and objective measurements of the trustworthiness of this technology. This will help the information society to choose the right IoT devices and services. We have to prevent efforts to compromise human agency with the IoT through consent fatigue or 'black-boxed' output by technical, social and legal means.

Ethical Concerns of IoT Technology

Although IoT technology is in the early stages of adoption in various applications across industry verticals, we are also seeing increased concerns among users about the security and privacy of data collected, stored and processed for smart-service

---

* Pasquale, F. (2015). *The black box society: The secret algorithms that control money and information.* Cambridge, MA: Harvard University Press.

offerings. Professor J. Zittrain* raises the following critical questions about AI ethics that are very significant for AI-enabled IoT applications:

■ "When does being able to offload thinking and decision making to an automated process enhance our freedoms, and when does it constrain them?"
■ "Under what circumstances could something be autonomy-enhancing for individuals, while constraining for society at large, and vice-versa?"

The key ethical concerns about the IoT that are gaining momentum are:

■ *Openness and transparency in design.* The users of IoT devices and services do not want a 'black boxed' output. They want to experience the various dimensions of the philosophy of IoT technology that are applicable for the smart device or service catered to them by product and service vendors. This is a basic need to build user trust in this technology.
■ *Contextual integrity and degree of mediation.* IoT users are concerned about personal data gathering by smart devices and services, often in the guise of beneficence. To address such concerns, users should be provided with any necessary information about contextual metadata that is being collected, processed and shared in order to produce the smart output. Vendors have to practice 'principled performance'† in order to confine the degree of mediation on contextual data only to the limit defined for the service through user consent and without causing any tangible or intangible harm to the user.
■ *Invisibility of data transactions in a distributed system of systems.* As smart systems integrate with each other due to the need for data transaction in the smart service ecosystem, output from one smart system will feed to other smart systems, creating dependent and interdependent smart systems, as in a smart city. In such IoT service offerings, care has to be taken to remove opacity in the movement of users' personal data across smart system boundaries.
■ *Trust permeability in interfaces between humans and the IoT.* As adoption of IoT devices and smart services increases, we will see enhanced interaction between smart devices, smart services and humans in a variety of combinations and degrees of confluence.

---

* Zittrain, J. (2017, May). Some starting questions around pervasive autonomous systems. Retrieved from https://medium.com/berkman-klein-center/some-starting-questions-around -pervasiveautonomous-systems-277b32aaa015
† Principled Performance®. (n.d.). Retrieved from https://www.oceg.org/about/what-is-principled -performance/

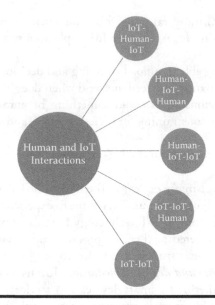

Figure 3.4 Five types of interactions between Human and the IoT.

These interactions can be segregated into five types, as shown in Figure 3.4 and explained below.

i. **IoT-Human-IoT**: This is a case of embodiment, whereby an implanted IoT device measures a biological parameter and instructs another IoT device for action, autonomously or through human mediation. For example, a smart glucometer measures the blood-glucose level of a user, and a smart insulin pump is instructed to pump out a measured volume of insulin into the blood-stream, either by the user or autonomously based on a defined algorithm in the smart device.

ii. **Human-IoT-Human**: In such interactions, humans define specific param-eters in the IoT device as needed, and then based on contextual alerts, take necessary actions. We find such examples in the smart-security scenario where a home owner works with their local police to fit a smart camera inside and outside their home to feed condition-based alerts to the police station to prompt necessary action.

iii. **Human-IoT-IoT**: Such interactions can occur when a human remotely inter-acts with a smart-home air conditioner for rapid cooling to a defined tem-perature and the smart home system talks to the smart grid system requesting electric power for the requested service.

iv. **IoT-IoT-Human**: In this interaction, two IoT devices interact with each other to generate a service, and the second IoT device interacts with the human user. For example, a smart refrigerator can scan the vegetable basket to identify

missing items that the owner has consumed, and can order these online to be
sent by a smart drone that confirms the order and dispatch address with the
refrigerator owner before delivery.

v. **IoT-IoT**: This is a classic case of delegating human autonomy and agency
to 'Things'. When mediated, these smart devices act on the behalf of the
mediator, which can be another smart device. New examples of such usage of
smart devices are coming up, as in smart-city interdependent systems where
smart transport-management systems identify city traffic jams and accordingly instruct the smart traffic system to divert traffic flows from clogged city
intersections.

Aside from (v), the other four IoT interaction scenarios depict a hermeneutic
relation between humans and 'Things', where the things provide a representation of
reality to be interpreted by humans. Profiling humans through mediations on their
smart things can lead to a breach of trust and degraded confidence among users.

To ensure trustworthiness in this emerging technology we have to ensure that
humans are not completely removed from the designing phase and are not sub-
jected only to end results of 'Things'. For developing a justified true belief in IoT
technology, the actions of 'Things' should be justified with an option to evaluate
the output.

As the IoT penetrates deeper in the societal fabric, creating social trust around
epistemic trust will be a challenge that needs to be addressed through enhancing the digital knowledge of humans, appropriate regulations and standardization
efforts. According to professor U. Gasser, "a governance approach to AI-based technologies embraces and activates a variety of modes of regulation, including technology, social norms, markets and law."*

We have to develop a reputation system for IoT devices in order to provide
explicit qualitative and objective benchmarks for trustworthiness. This will help the
information society to choose the right IoT devices and to enhance social awareness
about the expected outcomes of using these smart devices and services.

If humans rely on informed consent alone in their IoT usage, and if we are not
bounded by rationality to judge the output of IoT systems, then this can lead to
a compromising of human agency. We have to apply appropriate guidance on the
design and usage of IoT technology through technical, social, legal and regulatory means in order to prevent IoT devices and smart services from compromising
human agency, which can result from any kind of consent fatigue or 'black-boxed'
output.

Opacity of IoT technology by device manufacturers and smart service providers
can lead to the obfuscation of functionalities and result in knowledge dissonance
among users, in which users find that what they know about the smart device or

---

* Gasser, U. (2017, June). AI and the law: Setting the stage. Retrieved from https://medium.com
/berkman-klein-center/aiand-the-law-setting-the-stage-48516fda1b11

service is different from what they experience. Also, users can have variations in their knowledge about IoT technology due to epistemic capabilities and limited access to know-how. Such dissonances can be eliminated by preventing blindness to profiling algorithms and IoT functionalities through accountable algorithm designs, making 'Things' open for inspection and verification, an ethics-by-design approach, development of a universally accepted conformance system and a governance mechanism with appropriate regulatory measures. These efforts are necessary to develop user trust. We have to ensure that users develop epistemic capabilities of IoT technology through education, training, self-help portals, knowledge exchange in society and an appropriate grievance redress system. An ethical design of smart systems considering the various dimensions of philosophy of information discussed in this chapter will help us to realize the potential of this technology.

Suggested Reading

Britton, R. (2004). Subjectivity, objectivity, and triangular space. *The Psychoanalytic Quarterly, 73*(1), 47–61. Retrieved from http://onlinelibrary.wiley.com/doi/10.1002/j.2167-4086.2004.tb00152.x/full

Chaudhuri, A. (2015). Address security and privacy concerns to fully tap into IoT's potential. Retrieved from http://www.tcs.com/SiteCollectionDocuments/White%20Papers/Address-Security-rivacy-Concerns-Fully-Tap-IoT-Potential-1015-1.pdf

Chaudhuri, A. (2016). Internet of things data protection and privacy in the era of the General Data Protection Regulation. *Journal of Data Protection and Privacy, 1*(1), 64–75. Retrieved from http://www.ingentaconnect.com/content/hsp/jdpp/2016/00000001/00000001/art00009?crawler=true

Curvelo, P., Guimarães Pereira, A., Tallacchini, M., Rizza, C., Ghezzi, A., Vesnic-Alujevic, L., Breitteger, M., & Boucher, P. (2014). The constitution of the hybrid world: How ICT's are transforming our received notions of humanness. *JRC87274 (Scientific and Policy Reports)*. Retrieved from https://www.researchgate.net/profile/Caroline_Rizza/publication/264310236_The_constitution_of_the_hybrid_world_How_ICT's_are_transforming_our_received_notions_of_humanness/links/54ea56f20cf25ba91c82f888/The-constitution-of-the-hybrid-world-How-ICTs-are-transforming-our-received-notions-of-humanness.pdf

Floridi, L. (2002). On the intrinsic value of information objects and the infosphere. *Ethics and Information Technology, 4*(4), 287–304. Retrieved from https://link.springer.com/article/10.1023%2FA%3A1021342422699?LI=true

Floridi, L. (2002). What is the philosophy of information? *Metaphilosophy, 33*(1–2), 123–145. Retrieved from http://onlinelibrary.wiley.com/doi/10.1111/1467-9973.00221/pdf

Floridi, L. (2017). The logic of design as a conceptual logic of information. *Minds and Machines, 27*(3), 495–519. Retrieved from https://link.springer.com/article/10.1007/s11023-017-9438-1

Freiman, O. (2014). Towards the epistemology of the internet of things: Techno-epistemology and ethical considerations through the prism of trust. *International Review of Information Ethics, 22*, 6–22. Retrieved from https://philpapers.org/archive/FRETTE.pdf

Lindley, J.G., Coulton, P., & Cooper, R. (2017). Why the internet of things needs object orientated ontology. *The Design Journal, 20*(sup 1). Retrieved from https://www.researchgate.net/profile/Paul_Coulton/publication/315836994_Why_the_Internet_of_Things_needs_Object_Orientated_Ontology/links/58eb419c0f7e9b978f84adfc/Why-the-Internet-of-Things-needs-Object-Orientated-Ontology.pdf

Mittelstadt, B. (2017). Designing the health-related internet of things: Ethical principles and guidelines. *Information 8*(3), 77. Retrieved from http://www.mdpi.com/2078-2489/8/3/77/htm

Rogerson, S. (2017, July). Coding ethics into technology. *H&C News*. Retrieved from http://hncnews.com/coding-ethics-technology?lipi=urn%3Ali%3Apage%3Ad_flagship3_feed%3B%2F4VWfaArTUKMLc2vnUCIpQ%3D%3D

Searle, J.R. (1998). How to study consciousness scientifically. *Philosophical Transactions of the Royal Society of London B: Biological Sciences, 353*(1377), 1935–1942. Retrieved from http://rstb.royalsocietypublishing.org/content/353/1377/1935.full.pdf

Searle, J.R. (2006). Social ontology: Some basic principles. *Anthropological Theory, 6*(1), 12–29. Retrieved from http://journals.sagepub.com/doi/abs/10.1177/1463499606061731

Waddell, K. (2017, May 1). The internet of things needs a code of ethics. *The Atlantic*. Retrieved from https://www.theatlantic.com/technology/archive/2017/05/internet-of-things-ethics/524802/

'FOR THINGS'

Chapter 4

Potential Applications of Internet of Things Technology

We are all now connected by the Internet, like neurons in a giant brain.

Stephen Hawking

After reading this chapter you will be able to:

- Understand the potential use of the IoT in the domains of healthcare, the home, energy, automotives and retail
- Interpret the opportunity of enabling smart products and services in the above domains
- Get insight about how data can be gathered from various contexts and analyzed to provide smart services
- Know how the IoT and robotics are combined to provide smart offerings.

Introduction

The potential of IoT applications has been discussed in brief in Section I—'Of Things'. In Section II—'for Things', we will see how the IoT's potential is being unfolded in various domains. Some potential IoT-enabled products and services in healthcare, retail, automotives, energy and home appliances have been discussed here to understand the diversity of this technology in applications and to perceive the velocity of progress. This is not an exhaustive display of IoT applications, but only a view of the tip of the iceberg. New smart products and services are being conceived by businesses, governments and enthusiasts across the globe and across domains in various forms to realize the benefits of IoT technology.

Smart Healthcare

The Internet of healthcare Things, smart healthcare or healthcare IoT—all these are new terminologies for the emerging healthcare products and services with IoT integration. The healthcare industry is adopting the IoT at a rapid pace, and market analysis reports for healthcare IoT predict massive growth in next five to eight years. According to predictions, the healthcare IoT market can grow to $160+ billion by 2020.[*] According to McKinsey, the smart healthcare market can reach a size of up to $1.6 Trillion by 2025.[†]

Some of the potential IoT applications in healthcare that have been designed in recent times are discussed here. This is not exhaustive, as it is an emerging field with immense opportunities for exploration.

Connected medical devices. These have ready-to-use features like remotely accessing other devices and interacting with those devices and humans to monitor, collect, and analyze specific health data or transmit the data to other devices for storage and analysis, and to notify other devices, users or health practitioners about events. Example: smart glucose monitor.

Wearable healthcare. These can come in various forms like patches or bands and are used for monitoring of specific health data or medical conditions. Example: biosensor-embedded healthcare patch for health monitoring of patients, fitness trackers and blood pressure and temperature-monitoring devices.

Smart hospital monitoring and medical equipment maintenance. Some of the smart applications that we are seeing modernizing the hospitals embracing IoT are smart monitoring of patients and their location identification, remote patient monitoring, patient criticality-alert management systems, remote drug delivery, app-based patient data gathering and health record management, smart beds, remotely locating medical equipment, cloud-connected medical equipment to track assets and for maintenance, and pharmacy inventory management with RFID tags.

---

[*] IoT in healthcare market to reach $163.24 billion by 2020 - rise in investments to implement healthcare IoT solutions. (2017). Retrieved from https://www.prnewswire.com/news-releases /iot-in-healthcare-market-to-reach-16324-billion-by-2020---rise-in-investments-to-implement -healthcare-iot-solutions---research-and-markets-300452141.html

[†] Manyika, J., Chui, M., Bisson, P., Woetzel, J., Dobbs, R., Bughin, J., & Aharon, D. (n.d.). Unlocking the potential of the internet of things. Retrieved from https://www.mckinsey .com/business-functions/digital-mckinsey/our-insights/the-internet-of-things-the-value -of-digitizing-the-physical-world

Smart medical ingestibles. A recent advancement in this branch of the health-care IoT is the development of the 'smart pill' that has edible sensors to monitor adherence to prescribed intake of pills at different times of the day, which specifically suits the elderly with memory obstacles.

Smart cognitive healthcare. Smart monitoring devices are being developed for addressing cognitive impairment and dementia among the elderly, monitoring Parkinson's and assessing depression. These can function as early-warning predictive systems that analyze health data, make predictions and provide advice to doctors to make decisions on the course of treatment.

Health 'rob-IoT-ics'. New applications are being designed in healthcare utilizing the benefits of robotics and the IoT, hence the name 'health rob-IoT-ics' used here to aptly define this space. Not long from now, we will have robotic assistants that will know when we are stressed or tired or developing some medical criticality and will act in accordance to inform family members, a doctor and an ambulance service to prevent fatalities. Cloud-connected robotic personal assistants that use AI to personalize the interaction with human users can be helpful for the elderly and patients. Many new application domains are being explored that will gradually come into prominence in the coming years.

We will explore some of these smart healthcare products and services to understand the impact and opportunities being created by IoT technology in the healthcare domain.

The SmartPlate for a Cognitive Approach to Heart Health

Fitly has created a new category in healthcare service termed 'culinary medicine'. Inspired by Anthony Ortiz, founder and CEO, Fitly's team has designed and created the SmartPlate,* the world's first intelligent nutrition platform. It can instantly analyze a user's entire meal with accuracy.

As shown in Figure 4.1, the SmartPlate "uses advanced photo recognition and AI technology to identify, analyze, and track everything"† that the user eats. It can "track macronutrients (protein, carb, fat) and micronutrients (fiber, sugar, sodium) accurately and easily."‡

Based on dietary needs and doctor advice, the SmartPlate can help to "analyze and adjust meals accurately, easily and quickly."§ The SmartPlate app has over 400,000 food products that can be easily scanned for instant tracking and analyzing portions, caloric content and related data to help in weight management, building strength and power, maintaining hydration, controlling carbohydrates and for sustained energy and endurance.

---

* SmartPlate. (n.d.). Retrieved from https://www.getsmartplate.com/
† Ibid.
‡ Ibid.
§ Ibid.

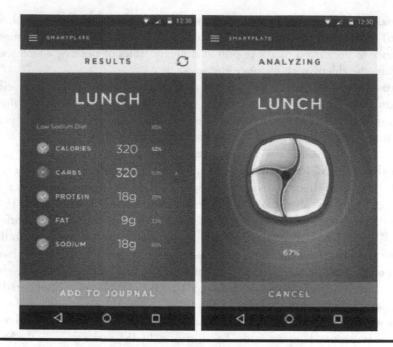

Figure 4.1   SmartPlate for a Cognitive Approach to Heart Health.

SmartPlate. (n.d.). Retreived from https://www.getsmartplate.com/

Kinsa Smart Thermometer

Kinsa* provides an FDA-cleared connected digital thermometer to measure and automatically store a patient's body temperature on a centralized database in the cloud.

Kinsa's app, as shown in Figure 4.2, tracks fever readings, symptoms, diagnoses, medication doses and other notes in a time-stamped log. The report is available for each member of a family. Patients can utilize the social community using the application to share health information with friends and groups.

Kinsa offers real-time guidance based on age and fever reading, so the user or family members know when to call a doctor and what to do next. The app-generated trend report for individual patients as well for the local population can help diagnose the spread of symptoms for necessary action.

---

* Kinsa. (n.d.). Retrieved from https://www.kinsahealth.com/

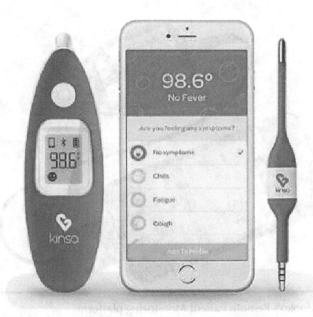

Figure 4.2 Kinsa Smart Thermometer.

Kinsa. (n.d.). Retreived from https://www.kinsahealth.com/

Glooko's Remote Patient Monitoring for Diabetes

Glooko* (Figure 4.3) provides an FDA-cleared and HIPAA-compliant diabetes management platform by remotely accessing a patient's diabetes-related health data from 50+ meters away, with insulin pump and continuous glucose monitoring (CGM), synced from a patient's iPhone or Android device. It enables patients and their healthcare team to see all their diabetes data in a central location. Glooko is a device-agnostic solution and is compatible with the majority of diabetes devices on the market. It helps to analyze a patient's daily intake of carbohydrates, food, insulin and medication next to blood glucose levels. It also provides insights for care adjustments and identifies at-risk patients for timely and efficient care.

Once patients sync their health data to the Glooko platform, it can be securely accessed from any web browser via the Glooko web app or via the Glooko mobile app. Patients can also grant access to a provider or group of providers, enabling them to securely access the health data from any web browser to provide remote support.

---

* Remote Patient Monitoring. (n.d.). Retrieved from https://www.glooko.com/landing/remote -monitoring/

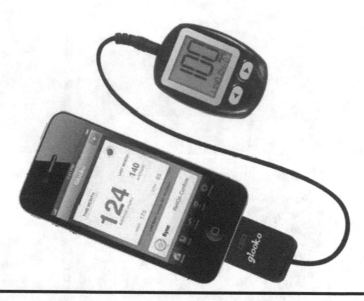

Figure 4.3 Glooko's Remote Patient Monitoring platform.

Remote Patient Monitoring. (n.d.). Retrieved from https://www.glooko.com/landing
/remote-monitoring/

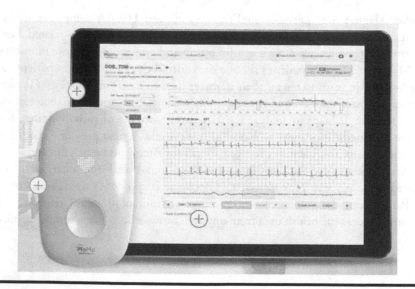

Figure 4.4 InfoBionic's smart healthcare device MoMe® Kardia.

From MoMe® Kardia. (n.d.). Retrieved from https://infobionic.com/

InfoBionic's MoMe® Kardia for Remote Cardiac Arrhythmia Detection and Monitoring

The FDA-cleared, ergonomically optimized MoMe® Kardia* device from InfoBionic has been designed to help detect cardiac arrhythmias in patients by sensing ECG, respiration, and motion (Figure 4.4).

MoMe transmits data via cellular means to a cloud-based platform where the data is analyzed. It works as a holter, event, and mobile cardiac telemetry (MCT) monitor. A physician can switch the device remotely to any one of three main monitoring modes based on the patient's cardiac symptoms and the need for different types of monitoring. The physician can access a patient's data via the web and also via iOS and the Android app. The device is lightweight and is easy for patients to wear as a necklace or belt attachment and is easy to use. The monitor continuously streams ECG and motion data for real-time review, and automated reports can be accessed for in-depth analysis of onset, offset, and other supplementary data in order to verify events on a 24/7-basis from any device through a HIPAA-compliant portal. Remote transition of the monitor is possible between holter, event, and MCT, as per a patient's need. All streaming data is verified via algorithm analysis in the cloud and is validated by QA professionals.

Animas® Vibe® Insulin Pump

Animas Corporation, from the Johnson & Johnson Diabetes Care Companies (JJDCC), has developed the FDA-approved OneTouch Vibe™ Plus Insulin Pump and Continuous Glucose Monitoring (CGM) System† for the treatment of patients aged two and older living with diabetes (Figure 4.5).

The insulin pump is integrated with Dexcom G5® mobile CGM technology, which includes the Dexcom G5® mobile sensor and transmitter. This system enables patients to see their glucose reading any time using the Dexcom G5® app on their smart phone, and delivers the precise amount of insulin needed by the body from the pump.

The Dexcom G5® transmitter collects blood glucose readings from the Dexcom sensor and wirelessly sends them to the patient's OneTouch Vibe™ Plus insulin pump screen and compatible smart device using the Dexcom G5® mobile system and app. Glucose data can also be shared with up to five people by utilizing the Dexcom Follow app.

The Animas® insulin pump contains a cartridge filled with rapid-acting insulin. It has a screen and buttons for programming the pump's internal computer. A precise

---

* InfoBionic. (n.d.). Retrieved from https://infobionic.com/
† The Animas®Vibe® Insulin Pump & CGM System. (n.d.). Retrieved from https://www.animas .com/diabetes-insulin-pump-and-blood-glucose-meter/animas-vibe-insulin-pump

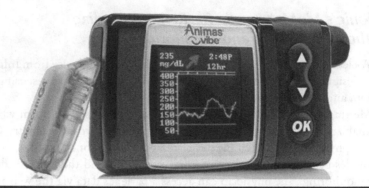

Figure 4.5 OneTouch Vibe™ Plus Insulin Pump and Continuous Glucose Monitoring CGM system.

From Animas® Vibe® System. (n.d.). Retrieved from https://www.animas.com/diabetes-insulin-pump-and-blood-glucose-meter/animas-vibe-insulin-pump

motor pushes insulin from the cartridge through tubing and an infusion set into the patient's body. The pump delivers two kinds of insulin: basal insulin and bolus insulin.

Patients and caregivers can access CGM data whenever and wherever it is most convenient for them in order to make informed diabetes-management decisions. For child patients, the parents and caregivers can continuously access their children's blood glucose in order to make the right treatment choices.

Apple Health

The Apple Health* app (Figure 4.6) consolidates health and fitness data from iPhone, Apple Watch and third-party apps to view the user's health status and to track progress in one place. It has four distinct categories: activity, sleep, mindfulness and nutrition. Each of these category plays an important role in understanding the overall health of the user.

The 'Activity' category of the Apple Health app provides information regarding how much the users moves. This information is obtained from a combination of activity data from the user's iPhone like the number of steps and distance travelled, with metrics from third-party fitness apps. The Apple Watch also automatically records simple, meaningful kinds of movement, like how often the user stands, how much he/she exercises and the all-day calorie burn.

Sleep data and sleep patterns can be tracked using the 'Bedtime' tab in the Clock app. This helps to establish a target bedtime and wake time and provides visuals of sleep patterns. These data are fed to the Health App, and combined with data from third-party sleep apps, the user can plan a healthier sleep routine. Time spent being mindful can also be tracked through deep breathing exercises and stress-relieving

---

* Apple. (n.d.). Retrieved from https://www.apple.com/in/ios/health/

Figure 4.6 Apple Health app.

From Apple Health app. (n.d.). Retrieved from https://www.apple.com/in/ios /health/

meditations performed. This adds to the health data on the Apple Health app. The Apple Health app collects important nutritional metrics as needed by the user, like carbohydrate intake, daily calorie intake and so on, and helps to manage individual nutrition goals. It also keeps tab on blood pressure, blood glucose, weight and other data as needed by the user. All data from Apple Health app can be backed up to iCloud and is encrypted while in transit and at rest.

Apple Health also provides access to ResearchKit and CareKit tools to help users participate in medical research with their data and manage their health.

Google Fit

The Google Fit* app (Figure 4.7) uses sensors built into users' mobile devices to effortlessly track fitness activities like walking, cycling and running, and fitness goals and body weight management progress over a defined period of time. Users can choose from a list of 120+ fitness activities to track.

The Fit App can be downloaded from the Google Play Store and is also preloaded on Android Wear watches. It provides instant insights with real-time statistics of

---

* Google Fit. (n.d.). Retrieved from https://www.google.com/fit/

Figure 4.7   Google Fit App.

From Google Fit - Fitness Tracking. (n.d.). Retrieved from https://play.google.com /store/apps/details?id=com.google.android.apps.fitness&hl=en

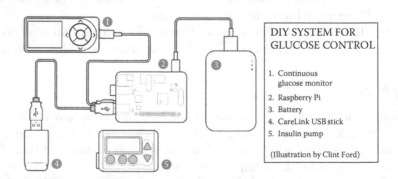

Figure 4.8   Smart System for Glucose Control.

From Rosenblum, A. (2015, December). This couple is hacking the insulin pump. Retrieved from http://www.popsci.com/hacker-medicine

activities performed. Google Fit records the user's speed, pace, route, elevation, and other relevant parameters to help the user stay motivated and on track. It also aggregates information from other apps to track fitness, nutrition, sleep, and body weight. Users can now automatically measure and track their heart rate throughout the day using the app.

The Open Artificial Pancreas System

The Open Artificial Pancreas System (OpenAPS) project,* developed by Dana Lewis and Scott Leibrand, is an open and transparent effort to make safe and effective basic artificial pancreas system technology widely available to quickly improve and save as many lives as possible and reduce the burden of Type 1 diabetes.

OpenAPS has been designed to automatically adjust an insulin pump's insulin delivery to keep blood glucose in a safe range at all times. OpenAPS does this by communicating with an insulin pump to obtain details of all recent insulin dosing (basal and bolus) by communicating with a CGM to obtain current and recent blood glucose estimates, and by issuing commands to the insulin pump to adjust insulin dosing as needed.

As shown in Figure 4.8, is designed to use existing approved medical devices, commodity hardware, and open-source software. The design has considered the primary factors of safety, understandability, and interoperability with existing treatment approaches and existing devices.

The OpenAPS dosing algorithm ensures that the decision it makes is the safest one possible with the information available at the time. By ensuring that all available information, including blood glucose level and trend information and insulin dosing history, is used in determining all insulin dosing decisions, OpenAPS can safely mitigate high blood-sugar levels while minimizing hypoglycemia risk.

OpenAPS is designed to simply and safely fall back to the patient's pre-programmed basal therapy whenever it receives conflicting information about what the appropriate course of action is or when required information is missing.

OpenAPS further ensures safety by falling back to traditional "low glucose suspend" behavior when current blood glucose is below a configured threshold and falling or not rising fast enough.

The Smart Pill—Medical Ingestible Sensors

Proteus Digital Health has developed Proteus Discover,† a smart system for patients, to easily monitor their medication-taking patterns on their mobile device (Figure 4.9). Proteus Discover is comprised of ingestible sensors, a small wearable sensor patch, an application on a mobile device and a provider portal.

---

* OpenAPS. (n.d.). Retrieved from https://openaps.org/
† Proteus. (n.d.). Retrieved from http://www.proteus.com/

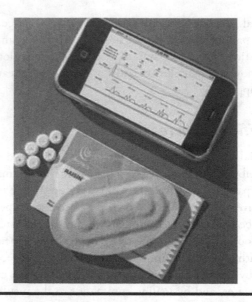

Figure 4.9   Smart pill.

From Aspler, S. (2014, September). How ingestible sensors and smart pills will revolutionize healthcare. Retrieved from https://www.marsdd.com/news-and-insights/ingestibles-smart-pills-revolutionize-healthcare/

A patient can activate Proteus Discover by taking prescribed medication with the ingestible sensor. When the ingestible sensor reaches the stomach, it transmits a signal to the patch worn on the torso, and a digital record is sent to the patient's medical device and the Proteus Cloud, which can be accessed by healthcare providers and caregivers with the patient's permission.

The ingestible sensor or smart pill is made of elements found in a typical diet and is FDA approved. The patch is an FDA-approved sensor that tracks medication taking and health parameters, and sends data to the Discover app. The 'Discover Portal' provides insights from patients' data and allows physicians to break down the data for further analysis.

Smart Wearables for Patients with Major Depressive Disorder (MDD)

Takeda Pharmaceuticals U.S.A. and Cognition Kit Limited are collaborating to pilot the use of a specially designed app on an Apple Watch wearable to monitor and assess cognitive function in patients with major depressive disorder (MDD).*

---

* Takeda and cognition kit partner to pilot wearables in patients with major depressive disorder (mdd). (2017, February 24). Retrieved from http://www.cambridgecognition.com/news/entry/takeda-cognition-kit-partner-pilot-wearables-patients-major-depression

Figure 4.10   STANLEY Healthcare's IoT-enabled RTLS badge.

From T14 Patient and Staff Badges. (n.d.). Retrieved from https://www.stanley healthcare.com/sites/stanleyhealthcare.com/files/documents/T14%20Badge%20 Data%20Sheet_0.pdf

MDD is the leading cause of disability worldwide, and cognitive problems are common in major depression. The Cognition Kit app will assess the pattern of cognitive symptoms in patients with MDD. It will advance patient assessment and monitoring in everyday life to help maximize patient engagement and potential treatment.

The app will collect real-time passive and active high-frequency mental health data with the help of a smart wearable. The participants in this study will wear an Apple Watch loaded with the Cognition App which will passively collect physical-activity data all day. At different times of the day there will be three separate prompts reminding the patients to complete cognitive and mood assessments.

This study will evaluate feasibility and compliance and will try to understand how measures of mood and cognition on wearable technology compare to more traditional neuropsychological testing and patient-reported assessments.

AeroScout T14 Staff and Patient Badge

STANLEY Healthcare's AeroScout t14* (Figure 4.10) is a Wi-Fi staff and patient badge for use with its AeroScout patient flow, nurse call automation, hand-hygiene compliance monitoring and MyCall staff protection solutions. AeroScout T14 is an IoT-enabled real-time location system (RTLS). The rechargeable battery-powered badge can be worn by staff members using a T14 badge cradle or can be secured to a patient's wrist using standard hospital bands. Bi-directional communication enables the badge to be programmed over the air. Bi-directional communication uses WPA2 personal security.

---

* AeroScout t14 staff badge. (n.d.). Retrieved from https://www.stanleyhealthcare.com/products /t14-staff-badge

The T14 badge utilizes lightweight beaconing communication (for standard messages) and bi-directional Wi-Fi communication with full network association and authentication.

For the MyCall staff protection solution, the badges include a call button for staff to call for help for themselves or others. Different button-press patterns can be programmed to distinguish between distress alerts and a call for assistance with patient care.

When combined with STANLEY Healthcare Exciters, the T14 badge provides instant notification when a patient or staff member passes through an egress point, such as a gate, doorway or other tightly defined area.

The badge can be integrated with specific solutions as per the need, like patient flow management for clinics, patient flow for emergency departments, patient flow for operating rooms and wireless nurse call systems integration.

Smart Retail

Leading retailers across the globe are exploring the potential of IoT technology to usher in new experiences for customers and to create differentiation in retail-service models.

Some of the notable smart applications that are developing in the retail industry are smart dressing/fitting rooms, RFID-enabled inventory tracking, smart mirrors, smart shelves, smart advertising using beacons, intelligent vending machines and store screening robots. We will discuss these smart applications below to gather more insight.

Smart Tagging with RFID and Connected Clothes at ZARA

Zara, one of the world's largest fashion retail companies, has implemented RAIN RFID-based stock tracking, inventory management and replenishment across multiple countries to achieve operational agility.*

The stores have small data centers, and RFID data is utilized for accurate sales monitoring, inventory compilation in a short time, restock alerts in real time and improved customer service (Figure 4.11). With the implementation of RFID technology, store employees do not need to scan inventories item by item. RAIN RFID handheld readers help to complete inventory scanning with a small team in a much shorter time.

Smart Mirrors

Smart mirror interactive technology, as shown in Figure 4.12, aims to assist consumers with their purchasing decisions while they are in the fitting room. Consumers can locate alternative items using the smart mirror, which has bar code scanning feature for selecting items on the mirror. Smart mirrors can display consumer reviews on various items and can also provide style recommendations to help shoppers decide on the right outfit.

---

* Advanced Mobile Mobile Group. (2015, December 21). How Zara controls stock with rfid. Retrieved from http://www.advancedmobilegroup.com/blog/how-zara-controls-stock-with-rfid

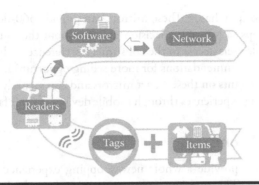

Figure 4.11 RAIN RFID-based stock tracking in a retail store.

What is rain?. (n.d.). Retrieved from http://rainrfid.org/about-rain/what-is-rain/

Retail giant Nordstrom has joined forces with eBay to use smart mirrors in their stores' dressing rooms. The luxury department store Neiman Marcus has utilized smart-mirror technology from MemoMi Labs. These smart mirrors are called memory mirrors. Using these mirrors, shoppers can try on clothes virtually to choose the right design, fit and color. These mirrors can be controlled by iPad, and shoppers can instruct these mirrors to save their images on them. These mirrors help to compare various outfits side by side, and clothing colors can be digitally changed to assist in

Figure 4.12 Smart shopping using a smart mirror.

Kothari, A. (2015, January 14). Effortless shopping with memo mi's "memory mirror". Retrieved from http://luxurylaunches.com/gadgets/effortless-shopping-with-memo-mis-memory-mirror.php

decision making for purchases. These mirrors can provide 360-degree views to the shopper, which helps in informed decision making about the right fit and color of clothing. The retailer gets access to the consumer's preferences, which helps them to send personalized recommendations for more selling opportunities. Consumers can create their own accounts on these smart mirrors and can securely store or send videos of their shopping experiences through mobile devices to their family and friends.

Smart Fitting Rooms

Smart fitting rooms provide a whole new shopping experience to consumers in fashion retail stores (Figure 4.13). Accenture, Avenade and Microsoft have prototyped a connected fitting room based on RFID technology. It tracks the items being brought into the fitting room by the shopper, and these items are up on the screen when the shopper enters the room to try out the chosen items.

By selecting different items and size and color combinations on the screen, the shopper does not need to go and collect these items again outside the fitting room.

Figure 4.13 Smart fitting room.

Wilson, M. (2014, June 11). Microsoft's smart fitting room is like a robo-shop clerk. Retrieved from https://www.fastcodesign.com/3031689/microsofts-smart -fitting-room-is-like-a-robo-shop-clerk

Instead, an alert is sent to the store attendant on their mobile device, who then brings the items to the shopper. The smart fitting room also provides outfit suggestions to the shopper based on the purchases of other people to help the shopper to decide on the right option. The smart fitting room also collects data from the shopper's buying habits and choices in order to provide a better shopping experience.

Smart Shelves

Smart shelves with digital-camera sensors and analytics technology are being used in retail stores like Kroger (Figure 4.14), the largest supermarket chain in the U.S. These are used to track shoppers' behavior toward products displayed on shelves with an aim to offer tailored pricing and suggestions on specific products and to call out products on the display screen based on items on the shoppers' mobile shopping lists. The mobile grocery list can also be connected with the smart shelf to alert the nearest location in the store where the item on the list is available, for the convenience of the shopper. The smart shelf can also light up when a potential shopper approaches a particular rack to pick an item on the shopper's list or something of his/her liking. Potential smart-shelf applications can be extended to electronic pricing boards to highlight the nutritional value of items when a shopper approaches a shelf.

Some smart-shelf technology can measure the shelf life of inventories and also identify which shelves are empty or are running low on products. Powershelf is a

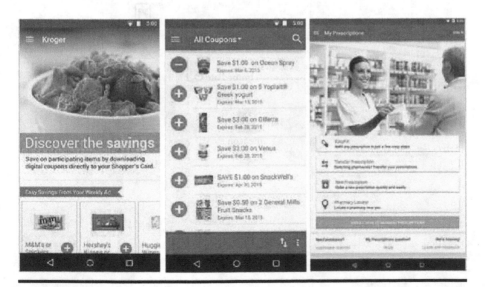

Figure 4.14   Kroger's smart shelf-connected app.

Anderson, G. (2015, October 07). What can smart shelf tech do for Kroger's business? Retrieved from https://www.retailwire.com/discussion/what-can-smart-shelf-tech-do-for-krogers-business/

smart-shelf technology with these features that runs on Microsoft Azure with Power BI dashboard. Supermarket chain Giant Eagle has reduced out-of-stock replenishment time by two-thirds and has cut out-of-stock SKUs by half using Powershelf.

Smart Beacons

Smart beacons with a proximity-detection feature have gained popularity in the retail industry for customer engagement and targeted marketing. Using smart beacons, retailers can attract more customers by sending notifications, special offers and discount coupons to shoppers in proximity to the retail outlets. These smart devices also help to build customer loyalty through targeted advertising and promotions to loyal customers based on their shopping habits and purchase history. Smart beacons can be enabled with Bluetooth Low Energy (BLE). For example, iBeacon from Apple, which was introduced in iOS7, leverages BLE. iBeacon defines regions within an identifier that can be used by multiple devices.

Fashion retailer Nordstrom Rack is piloting in-store beacon technology from Footmarks built on Microsoft Cloud. It aims to enhance customer experience and engagement in store. Shopkick has developed an app that works in tandem with

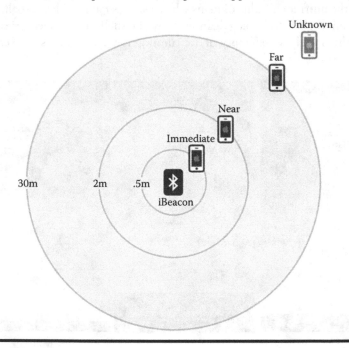

Figure 4.15 iBeacon.

Thompson, D. (n.d.). Introduction to iBeacon: Taking Bluetooth le for a test drive. Retrieved from http://beekn.net/2013/11/ibeacon-example-taking-bluetooth-le -for-a-test-drive/

Figure 4.16   Shelfie Robot.

Our new invention has them shouting in the aisles. (n.d.). Retrieved from https://lakeba.com/shelfie-robot-retail-getting-noticed/

a beacon product named ShopBeacon (Figure 4.15). The beacon will identify the right consumers walking past it in a particular location to feed offers to their mobile devices based on their preferences and past purchases.

Store-Scanning Robots

Microsoft Australia and Lakeba have partnered to develop Shelfie, the in-store robot, combining robotic technology with intelligent image capturing and data analytics running in the cloud (Figure 4.16). Shelfie, available as a robot or drone, can automatically travel inside the store and can scan shelves for real-time stock reporting using the Shelfie Dashboard, built on Microsoft Power BI.

This smart robot empowers retailers by identifying sales trends and improving store management and optimizing merchandise layout. Shelfie can check a supermarket's inventory within a few hours depending on store size. A new key performance indicator (KPI) has been developed, termed 'Shelfie Index', that provides a score between 0 and 100 to a store based on insights from stock and layout.

Intelligent Vending Machines

Mars Drinks, in collaboration with Neal Analytics, has applied remote-sensing technology to its vending machines and connected these to Microsoft Azure IoT Suite, the Cortana Intelligence Suite and Microsoft Power BI technology for predictive analysis of stock levels to understand consumer behavior and to apprehend

Figure 4.17   Smart vending machine.

KLIX®. (n.d.). Retrieved from http://klix.co.uk/vending-machines/klix-momentum

changes in demand based on influencing functions like weather or holidays (Figure 4.17). By applying the IoT, stock levels can be managed efficiently and revenue losses can be prevented.

Amazon Go

The physical retail store from Amazon, named Amazon Go, does not require a shopper to check out, and hence no time is wasted by the shopper at checkout queue. Amazon Go provides a 'Just Walk Out Shopping'* experience using the Amazon Go App that can be used for entering the store, taking necessary products and walking out (Figure 4.18). This is possible because Amazon Go can automatically detect the products taken by the shopper from the shelves or returned to the shelves at the store, and only those items taken are tracked in a virtual cart.

After the shopper leaves the store with selected products, the Amazon account of the shopper is charged accurately and a receipt is sent to the shopper. Amazon

---

* Amazon. (n.d.). Retrieved from https://www.amazon.com/b?node=16008589011

Figure 4.18   Amazon Go smart retail outlet.

Hardawar, D. (2017, December 5). Amazon Go is a grocery store with no checkout lines. Retrieved from https://www.engadget.com/2016/12/05/amazon-go-grocery -store/

Go uses sensors, computer vision, AI and related technologies to provide the smart retail experience. The first Amazon Go store started operating from Seattle, Washington.

Smart Home

Globally we are seeing an increased interest among customers for smart-home products and services in the segments of energy management and climate control, security and access control, smart lighting, smart entertainment, assisted living and home automation. According to McKinsey, IoT applications in the home can have an economic impact of $200 billion to $300 billion in 2025.* The smart home applications will be mostly in chore automation and security. The top motivators driving acceptance of smart homes are security and convenience.

Various types of interconnectivity and automation can be designed for smart-home solutions. For fully decentralized smart homes, the smart devices can function in autonomous mode and utilize the home network to connect to the Internet for transmitting data. Another alternative is to connect the smart devices locally like an intranet. A third option is to create a central hub or gateway for the smart devices.

---

* McKinsey & Company. (2015, June). The internet of things: Mapping the value behind the hype. Retrieved from https://www.mckinsey.com/~/media/McKinsey/Business%20Functions /McKinsey%20Digital/Our%20Insights/The%20Internet%20of%20Things%20The%20 value%20of%20digitizing%20the%20physical%20world/The-Internet-of-things-Mapping -the-value-beyond-the-hype.ashx

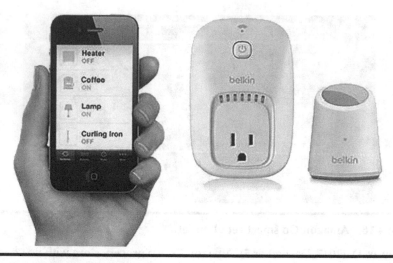

Figure 4.19   Belkin WeMo Home Automation device and app.

Reviews on Belkin WeMo home automation. (n.d.). Retrieved from http://www .domotics.sg/reviews-on-belkin-wemo-home-automation/

Belkin Wemo

The WeMo Home Automation* portfolio from Belkin has multiple smart products in its portfolio like the Wi-Fi Smart Dimmer, Mini Smart Plug, Smart Light Switch and Insight Smart Plug (Figure 4.19).

The WeMo Smart Dimmer sets room ambience and assists in controlling lights from anywhere. It connects with the home Wi-Fi network to provide wireless control of lights. WeMo Smart Dimmer will work seamlessly with Google Assistant and Amazon Alexa for hands-free voice control of the lights. It also helps to set up auto on-off schedules for any light as per user need. It can be paired with Nest's Thermostat for automotive control and also works with 'If This, Then That' free web-based service.

The WeMo Mini Smart Plug helps to control electronic appliances wirelessly from the phone or tablet using the home Wi-Fi network. It does not require any hub connectivity or subscription service. It also supports hand free assistance.

The WeMo Insight Smart Plug monitors home energy consumption from the phone or tablet. It provides real-time reports of device-wide energy consumption and energy cost.

---

* Belkin. (n.d.). Retrieved from http://www.belkin.com/in/Products/c/home-automation/

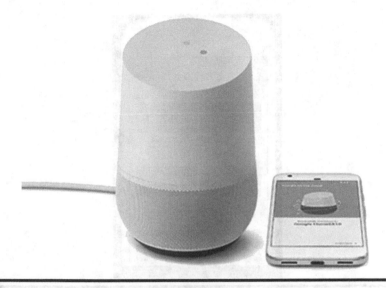

Figure 4.20   Google Home device and app.

Google Home. (n.d.). Retrieved from https://store.google.com/product/google
_home

Google Home

Google Home* is a smart, voice-activated speaker with hands-free help from Google Assistant (Figure 4.20). It is activated by the voice command 'OK Google' or 'Hey Google'. It supports multiple users and can distinguish individual voices for a personalized experience. With Google Home, users can make calls by voice command. If a particular contact has multiple numbers, then Google Assistant will read out the associated numbers of the contact to help the user choose the right number by voice command.

Some of the other features of Google Home are: getting answers from Google on various topics like weather, traffic, finance, sports and local business; managing personal tasks; daily planning; controlling home appliances and devices and entertainment features like Bluetooth audio, multi-room audio, music, news, podcasts, radio, speaker streaming and TV streaming. It can also order food online or book a cab as per the user's need. The entire top of the smart device is a touch surface to control the device. The content and features of Google Home can also be accessed from the Google Home app.

---

* Introducing Google Home. (n.d.). Retrieved from https://support.google.com/googlehome
  /answer/7029281?hl=en

Figure 4.21   Rachio Smart Sprinkler Controller app.

Rachio. (n.d.). Retrieved from http://rachio.com/how-it-works

Rachio Smart Sprinkler Controller

The Smart Sprinkler Controller from Rachio* (Figure 4.21) is an app-based Wi-Fi-enabled device that can monitor and adjust a sprinkler system from anywhere and anytime. It helps to control sprinklers remotely and water bills can be adjusted as per need, all from a smartphone. Rachio gathers local forecasts, soil and plant types, sun exposure and more information to create a customized watering schedule that saves the user both water and money.

The sprinkler controller skips watering before, during and after rain. Using the app the user can check and adjust sprinklers anytime and from anywhere to create customized zones. Rachio automatically calculates watering cycles to deliver the right amount of water, minimizing wastage through draining.

The user receives an alert when Rachio steps in to adjust schedules for rain, snow and seasonal changes. This smart device also integrates with connected-home platforms like Amazon Alexa, Google Assistant, Nest and other devices.

June Intelligent Oven

June Smart Oven is a smart countertop convection oven[†] with a built-in HD camera that allows the oven to know what is being cooked and suggests the best

---

* Rachio. (n.d.). Retrieved from http://rachio.com/how-it-works
[†] June. (n.d.). Retrieved from https://juneoven.com/

Figure 4.22   June Smart Oven app.

Bell, K. (2015, June 9). Smart oven uses image recognition to cook the perfect meal every time. Retrieved from http://mashable.com/2015/06/09/june-smart -oven/#epHESmpbdPq5

technique of cooking (Figure 4.22). This smart oven is fitted with a food thermometer that lets the oven inform the user when food is cooked. June has an iOS app that allows users to control their ovens from anywhere. Users can also remotely watch the food being cooked from a smartphone and can be notified when the meal is ready.

The oven's internal HD camera and food ID technology can automatically identify a variety of food. The oven selects the correct adaptive preset cook program to cook the food as per the user's preference, with no requirement to preheat. The food thermometer measures the internal temperature of the food to ensure that it is cooked exactly to the user's liking. The oven's software updates automatically and wirelessly to have the latest technology as it cooks. The June app can be paired with multiple devices to control the oven from anywhere. Customized alerts can also be set to communicate with the oven from a smart phone, watch or tablet.

Figure 4.23 Philips Hue personal wireless lighting system.

Hue white and color ambiance starter kit E27. (n.d.). Retrieved from https://www2 .meethue.com/en-in/p/hue-white-and-color-ambiance/8718696725405

Smart Lighting with Philips Hue

Hue* is a personal lighting system from Philips that allows users to easily control their lights' visual parameters to create the right ambience (Figure 4.23). Users can play with light by opting for variations from 16 million color choices and can sync up the smart light with music, games and movies.

Philips Hue can be connected with a range of different smart devices like Amazon Echo, Google Home, Apple Home Kit and other smart devices. The lights can be controlled by voice when connected with smart voice assistants. The favorite light settings can be saved and recalled with the tap of a finger whenever the user wants. The lights can be controlled remotely using the Philips Hue iOS and Android Apps.

Smart Security with LG Smart Security Wireless Camera

As shown in Figure 4.24, this wireless camera from LG[†] has a 5 megapixel (MP) image sensor that keeps watch over the user's home even when he/she is outside. Using the ADT Canopy App for LG Smart Security, the user can set up custom alerts as needed so that the user can be connected with appropriate alert notifications.

The 130-degree lens of the smart security camera and insight vision capability can capture wide angle views of a room to capture complete views of what is happening there. Video footage can be stored in the ADT Canopy cloud for retrieval and review as required. The app can connect other home automation devices for an integrated smart-home experience with Z-wave technology.

---

* Wireless and smart lighting by Philips | meet Hue. (n.d.). Retrieved from https://www2 .meethue.com/en-in

[†] LG Smart Security Wireless Camera. (n.d.). Retrieved from http://www.lg.com/us/home-security /lg-LHC5200WI

Figure 4.24 LG Smart Security Wireless Camera.

LG LHC5200W. (n.d.). Retrieved from http://www.lg.com/us/home-security/lg-LHC5200WI

Amazon Echo Dot

As shown in Figure 4.25, Amazon Echo Dot (2nd Generation)* is a "hands-free, voice controlled device that uses Alexa to play music, control smart home devices, make calls, send and receive messages, provide information, read the news, set alarms, read audiobooks from Audible, control Amazon Video on Fire TV, and more. When you want to use Echo Dot, just say the wake word 'Alexa' and Dot responds instantly."[†] It can connect to speakers or headphones through Bluetooth or a 3.5 mm stereo cable to play music. It can control lights, fans, TVs, switches, thermostats, garage doors, sprinklers, locks, and compatible connected devices from other vendors.

For hands-free control, the smart device can hear human voices from across a room with seven far-field microphones. The device uses echo spatial perception technology to respond to human voice commands intelligently. Users of Echo, Echo Dot, Echo Show or the Alexa App can talk to each other or send messages. Various actions are possible with Echo Dot, like switching on a lamp, dimming the lights, controlling room temperature and so on.

Connected Cars

Connected cars are fitted with mobile internet technology and sensors to gather contextual data, communicate remotely and control key functions remotely and wirelessly. Autonomous vehicles are an extended variety of connected cars with self-driving and robotic features. These cars are futuristic commuting means for

---

* Echo Dot (2nd Generation). (n.d.). Retrieved from https://www.amazon.com/Amazon-Echo-Dot-Portable-Bluetooth-Speaker-with-Alexa-Black/dp/B01DFKC2SO
[†] Ibid.

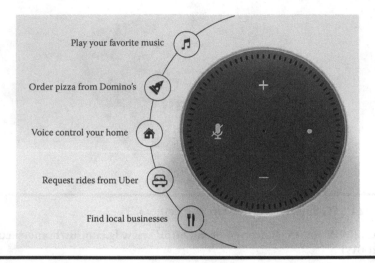

Figure 4.25   Amazon Echo Dot (2nd Generation).

Echo Dot (2nd Generation). (n.d.). Retrieved from https://www.amazon.com/Amazon-Echo-Dot-Portable-Bluetooth-Speaker-with-Alexa -Black/dp/B01DFKC2SO

a greener tomorrow. Some potential examples of connected car technology have been described here to provide a glimpse about the emerging smart automotive domain.

Tesla Connected Car

Tesla has incorporated internet technology in its vehicles for a digital driving experience. Tesla vehicles provide app-based remote climate control, voice-activated controls and other intelligent features.* All Tesla vehicles are fitted with hardware for full self-driving capability with safety features (Figure 4.26). Advanced sensor coverage has eight surround cameras and twelve ultrasonic sensors that detect hard and soft objects.

Tesla vehicles also have forward-facing radar that can see through heavy rain, fog, dust and cars ahead. A deep neural-network based vision processing tool named Tesla Vision processes vision data, server data and radar data on a powerful computer onboard. Tesla Vision deconstructs the car's environment at great levels of reliability. Tesla Autopilot can match speed to traffic conditions, can keep within driving lanes or automatically change lanes, transition to freeways, self-park in a parking spot and can be summoned to and from a garage. These vehicles can also have over-the-air software updates as needed.

---

* Taylor, J., (2016, August). Why your Tesla can become the ultimate 'connected car'. Retrieved from https://www.teslarati.com/tesla-model-s-x-connected-car/

Figure 4.26 Tesla connected car.

Taylor, J., (2016, August). Why your Tesla can become the ultimate 'connected car'. Retrieved from http://www.teslarati.com/tesla-model-s-x-connected-car/

Automobile Maps for Connected Cars

The global automobile company Toyota has grouped with Intel, Ericsson and Denso to create the Automotive Edge Computing Consortium,* which will build data pipes and collect and analyze Big Data to support intelligent driving, to create maps with real-time data and to provide cloud-based driving assistance for connected cars. Toyota has estimated that data volumes from vehicles and the cloud will reach 10 Exabytes per month by 2025.

Similar efforts have been made through a concerted effort by Daimler, BMW, Volkswagen and Here, the digital map maker (Figure 4.27). The three auto companies will contribute visual data to Here's systems. The data collected from brakes, windshield wipers, headlights, location systems, cameras and other sensors are to be utilized to create alerts on driver dashboards using Here's capabilities. Crowd-sourced data will help to create live maps and views of road conditions to aid the information systems of connected cars. The connected cars will perform like edge data centers constantly capturing, sharing and accessing live automotive data and road network data for informed decision-based actions.

---

* Ericsson. (2017, August). Industry leaders to form consortium for network and computing infrastructure of automotive big data. Retrieved from https://www.ericsson.com/en/news/2017/8/consortium-for-automotive-big-data

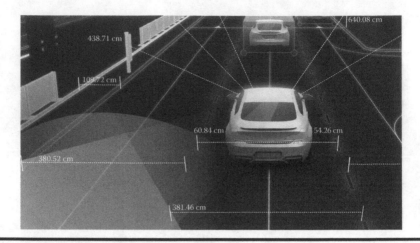

Figure 4.27   Here HD live map for automobiles.

Anon. (2017, February). Professors Building Self-Locating Autonomous Vehicles. Retrieved from https://www.fenderbender.com/articles/8581-professors-building -self-locating-autonomous-vehicles

Sensors to Reduce Car Crashes

Apple has designed Bluetooth sensors that can scan the surroundings of a car and update the driver's display to provide visuals of obstacles and passing vehicles to prevent car crashes. Apple has filed a patent* for this invention for commercial use.

Google's Waymo

Waymo,[†] Google's self-driving car project, aims to provide the owners of self-driving cars with valuable commute time that they can spend on their priority activities as they travel while the car handles the driving.

These self-driving cars (Figure 4.28) have sensors and software designed to detect pedestrians, cyclists, vehicles, road work and other relevant information from a distance and from all directions. These cars have AI-based capabilities to predict the behavior of all road users around them. Waymo relies on millions of miles of real-world driving experience to teach the cars how to navigate safely and comfortable through everyday traffic.

---

* Shukla, V. (2017, August). Patent hints at Apple Inc. (AAPL) car system that communicates with other vehicles. Retrieved from https://investorplace.com/2017/08/patent-hints-apple-car-system -communicates-with-other-vehicles-ggsyn/#.WkommtKGPIU
† Waymo. (n.d.). Retrieved from https://waymo.com/

Figure 4.28   Google's self-driving car project.
Waymo. (n.d.). Retrieved from https://waymo.com/journey/

Connected Car Data Platform

BMW and IBM are collaborating to collect and analyze data from connected cars (Figure 4.29). BMW will integrate its CarData* network with Bluemix, IBM's cloud infrastructure, and with access to Watson's IoT capabilities.

CarData can collect telematics data from connected cars with driver consent, like mileage, fuel levels, event data, component health data, service information, and so on, which will be encrypted and sent to BMW's secure servers. A SIM card will be utilized for data access and will be permanently installed in the vehicle. Third parties like automobile repair shops or insurance companies can access these data only after the data owner's consent. The car owner can decide on the third parties with whom these data can be shared based on specific service needs. The car-owner-authorized third parties will have access to only as much a car's data as is needed to provide their services, and they can utilize intelligence and cloud tools on the Bluemix platform to design customized offerings.

Remoto Connected-Car Platform

Remoto, shown in Figure 4.30 is an AI-enabled connected car platform[†] from Bright Box. Customers using the Remoto app can control various car features remotely.

---

* BMW. (n.d.). Retrieved from https://www.bmw.com/en/topics/fascination-bmw/connected-drive
  /bmw-cardata.html
[†] Remoto. (n.d.). Retrieved from https://remoto.com

Figure 4.29   Connected car data analytics.

BMW. (n.d.). Retrieved from https://www.bmw.com/en/topics/fascination-bmw/con nected-drive/bmw-cardata.html

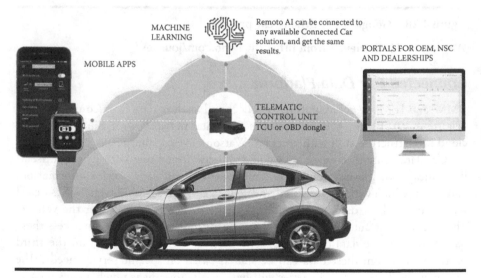

Figure 4.30   Remoto connected-car platform.

Remoto. (n.d.). Retrieved from https://remoto.com

The Remoto app-installed smartphone can be used as an I-Key to get access to the vehicle. In case of an accident, Remoto can automatically request help, or the car owner can seek road assistance from service providers using the app.

The app also helps to generate car diagnostic reports. Based on driving behavior, the AI-based app can make predictions for the next service visit. Gamification features on the app allow the user to save fuel, drive safely and to share experiences on social media. The app can also help to locate the nearest gas station when needed.

Figure 4.31   Glow smart-energy tracking device.
Horsey, J. (2017, August). Glow Home Smart Energy Tracker Can Help You Reduce
Your Bills (video). Retrieved from https://www.geeky-gadgets.com/glow-home-smart
-energy-tracker-09-08-2017/

Smart Energy

We are seeing bigger strides in the use of IoT in the smart-energy domain. Age-old
grid systems are being upgraded to smarter energy systems for quick mapping of
demand and planning for efficient supply of electricity to households and industries. Smart-energy devices and integrators are available for home use with a variety
of features. Some potential smart-energy initiatives are discussed below to highlight
the breadth of IoT applications in this domain.

Smart Energy Home Tracker

Glow* is a smart energy tracker that helps users understand how their homes are
expending energy (Figure 4.31). This smart device uses a magneto-resistive sensing
technology, and the sensor wirelessly measures the home's electric usage. The measured information is sent to an in-home display unit that changes color with the
home's energy usage.

The cross-platform Glow app provides home energy usage information that can
be accessed on the go. Notification alerts of abnormal usage can be configured on
the Glow app to prevent unnecessary energy wastage and save money. Glow also
estimates energy costs for the next bill based on current usage.

---

* Glow. (n.d.). Retrieved from https://meetglow.com/

Figure 4.32 IoT-enabled smart energy management in a smart city.

San Diego to deploy world's largest city-based 'internet of things' platform using smart streetlights. (2017, February 22). Retrieved from https://www.sandiego .gov/mayor/news/releases/san-diego-deploy-world%E2%80%99s-largest-city -based-%E2%80%98internet-things%E2%80%99-platform-using-smart

Smart Energy Initiatives in Smart Cities

The smart city of San Diego (Figure 4.32) has partnered with GE for upgrading its streetlights as part of a transformation initiative to reduce energy costs and to integrate streetlights with a connected digital network.* About 3000 smart sensors will be deployed in the city to create this digital network.

Barcelona, a smart city, has deployed about 19000 smart energy meters and more than 1100 LED streetlights with embedded sensors to monitor noise, weather and traffic. SIARQ, a Catalonian company, has implemented smart solar streetlights in Barcelona that have zero emissions and can incorporate various smart city sensors as per the need of the urban environment in order to create an intelligent sensory and information hub. Similar initiatives are also being implemented in other smart cities across the globe for a cleaner and greener living experience.

---

* San Diego to deploy world's largest city-based 'internet of things' platform using smart streetlights. (2017, February 22). Retrieved from https://www.sandiego.gov/mayor/news/releases/san-diego -deploy-world%E2%80%99s-largest-city-based-%E2%80%98internet-things%E2%80%99 -platform-using-smart

Figure 4.33 Ecobee smart thermostat.

Ecobee. (n.d.). Retrieved from https://www.ecobee.com/ecobee4/

Voice-Enabled Smart Thermostat

Ecobee has developed the next generation smart Wi-Fi thermostat,* which is voice enabled with built-in far-field voice technology and Amazon Alexa voice service (Figure 4.33). It has a room sensor to manage hot and cold spots. By a simple voice command like 'Alexa, I'm away', the user can lower the room temperature for energy savings. After placing this smart device in a room, it reads the temperature of the room and detects its occupancy. It can adjust the room temperature with total hands-free control.

The Ecobee mobile app helps adjust temperature and comfort settings on Android and iOS devices. It also generates energy reports on consumption details and provides information on heating and cooling equipment.

The Nest Learning Thermostat[†] from Google uses exclusive software algorithms to automatically lower air-conditioning costs by shutting off the compressor at the right time. It also learns information about a user's home, like how long it takes to warm up and how to make its heating and cooling system more efficient as the weather changes.

Smart Meter

The usage of smart meters (Figure 4.34) is gaining popularity, as these devices can measure the usage of electricity in a home and display it with the billing amount on a smart display or app-based dashboard. The smart meter can automatically send its reading to the energy supplier for accurate billing, removing estimation-based billing.

---

* Ecobee. (n.d.). Retrieved from https://www.ecobee.com/
† Nest. (n.d.). Retrieved from https://nest.com/thermostats/nest-learning-thermostat/overview/

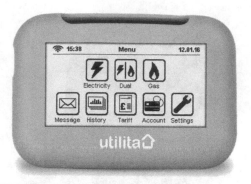

Figure 4.34   Smart meter.

Utilita Smart Meters. (n.d.). Retrieved from https://utilita.co.uk/smart-meters

Using a smart meter, the user gets energy usage reports in near real-time, along with a break-down of values of consumption by hour, week and month, with corresponding cost. It also helps users compare their energy usage with similar households in their region.

Smart Grid

The electric grid, comprising of a substation, transformers, a transmission network and meters, has been digitized and transformed into the 'smart grid'. According to International Energy Agency:

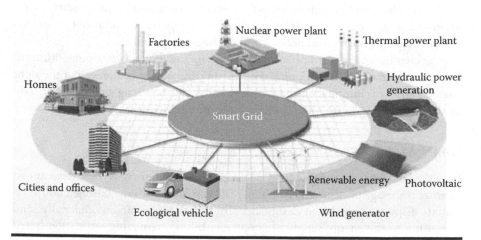

Figure 4.35   Smart grid dependent systems.

Lohrmann, D. (2017, March). Smart Grid Security: Is Trouble Coming? Retrieved from https://mitechnews.com/cyber-defense/smart-grid-security-trouble-coming/

A smart grid is an electricity network that uses digital and other advanced technologies to monitor and manage the transport of electricity from all generation sources to meet the varying electricity demands of end-users. Smart grids co-ordinate the needs and capabilities of all generators, grid operators, end-users and electricity market stakeholders to operate all parts of the system as efficiently as possible, minimizing costs and environmental impacts while maximizing system reliability, resilience and stability.*

With a smart grid, we can have instantaneous data about the volume of electricity generated, consumed, demanded and available to help with agile planning for production and distribution, as shown in Figure 4.35. Use of sensors, information technology and two-way communication between the utility provider and the consumers are prime features of a smart grid. Utilizing intelligence, automation and control, the smart grid will enable large-scale integration of renewable energy systems, real-time monitoring of grid operation and performance, reduced peak demand for efficient demand-side management, reduced energy wastage and early warning for blackouts.

As the world moves to an era of electric vehicles and smart cities, smart grids will play a crucial role as an enabler of this transformation. As per Frost & Sullivan, the global smart grid Industry will grow to $112.7 billion in 2025.† ABB sees the future of electricity in smart grids across the globe.‡ With smart grids, we can transform the archaic electric-grid infrastructure into a digitized version that is more reliable, available and efficient.

The government of India has initiated smart grid pilot projects§ across the country to provide direct benefits to consumers through real-time monitoring of energy consumption, automated outage management, faster restoration, improved reliability and better quality of electric power, contributing to economic and environmental health.

Remote Energy-Management Solutions

Organizations with expansive and geographically distributed sites face concerns regarding inefficient energy usage and an enlarged energy footprint due to lack of control and monitoring of energy usage to finer detail in all the sites. Applying the IoT, Big Data and analytics in smart energy solutions, we can monitor real-time electricity consumption with a centralized energy-management automated-alert system.

---

* IEA. (2011). Technology roadmap. Retrieved from https://www.iea.org/publications/freepub
  lications/publication/smartgrids_roadmap.pdf
† Frost & Sullivan. (2016, May 26). Future of the smart grid industry. Retrieved from http://
  www.frost.com/prod/servlet/resolvehost/mc2e
‡ ABB. (2008). When Grids Get Smart. Retrieved from http://thehill.com/sites/default/files/ABB
  _WhenGridsGetSmart_0.pdf
§ National Smart Grid Mission. (n.d.). Retrieved fromhttp://nsgm.gov.in/en/sg-projects

An award-winning example is the TCS Remote Energy Management Solution (REMS). It is:

> A smart, scalable, analytics-driven energy monitoring and management solution that helps organizations optimize their energy usage and enhance their efficiencies in operations of facilities. It provides real time visibility for organizations on their energy consumption.*
>
> REMS has been deployed in TCS in more than 100 of its facilities, which host more than 280,000 employees, across over 30 million square feet. It enables round-the-clock monitoring of energy consumption. TCS Connected Universe Platform (TCUP), an IoT based platform, serves as the backbone to the solution to manage information from separate sources like sensors, metering devices, and so on, over different communication protocols. REMS has enabled TCS' eco-sustainability vision to ensure continued business growth with no commensurate increase in the organization's power consumption.†

The IoT applications discussed abive provide an insight on the breadth and depth of the IoT's potential and applicability. Readers interested in knowing more about these applications can explore the available knowledge and the product or service brochures from the providers.

Suggested Reading

Alter, L. (2017, August). Glow is a cute and "smart energy tracker for your home". Retrieved from https://www.treehugger.com/gadgets/glow-cute-and-smart-energy-tracker-your-home.html

Animas Corporation. (n.d.). My insulin pump workbook. Retrieved from https://www.animas.com/sites/animas.com/files/pdf/ANCO_GEND_0317_0056%20My%20Insulin%20Pump%20Workbook.pdf

Apple CarPlay The ultimate co-pilot. (n.d.). Retrieved from https://www.apple.com/in/ios/carplay/

Aspler, S. (2014, September). How ingestible sensors and smart pills will revolutionize healthcare. Retrieved from https://www.marsdd.com/news-and-insights/ingestibles-smart-pills-revolutionize-healthcare/

Balakrishnan, A. (2017, August). Apple has an idea for car sensors that could drastically reduce crashes. Retrieved from https://www.cnbc.com/2017/08/17/apple-patent-shows-connected-car-sensors-v2v.html

---

* Tata Consultancy Services. (2017, February 8). TCS Remote energy management solution wins 2016 IoT award for connected building. Retrieved from https://www.tcs.com/tcs-remote-energy-management-solution-wins-2016-iot-award-connected-building

† Ibid.

BMW connected drive services. (n.d.). Retrieved from https://www.bmw.com/en/topics/fascination-bmw/connected-drive/bmw-cardata.html

Casey, S. (n.d.). San Diego to deploy world's largest city-based 'internet of things' platform using smart streetlights. Retrieved from http://cleantechsandiego.org/san-diego-deploy-worlds-largest-city-based-internet-things-platform-using-smart-streetlights/

Castle, L. (2017, June). Connected cars may soon become edge data centers. Retrieved from https://360.here.com/2017/06/01/connected-cars-may-soon-become-edge-data-centers/

Connected car and IoT automotive cloud services. (n.d.). Retrieved from https://www.kaaproject.org/automotive/

Full self-driving hardware on all cars. (n.d.). Retrieved from https://www.tesla.com/autopilot

Gartner says connected car production to grow rapidly over next five years. (2016, September). Retrieved from http://www.gartner.com/newsroom/id/3460018

Graziano, D. (2014, October). The complete guide to Google Fit. Retrieved from https://www.cnet.com/how-to/the-complete-guide-to-google-fit/

Hafezi, H., Robertson, T. L., Moon, G. D., Au-Yeung, K. Y., Zdeblick, M. J., & Savage, G. M. (2015). An ingestible sensor for measuring medication adherence. *IEEE Transactions on Biomedical Engineering, 62*(1), 99-109.

Helping automakers drive innovation through connected cars. (2017, January). Retrieved from https://blogs.microsoft.com/iot/2017/01/05/helping-automakers-drive-innovation-through-connected-cars/

How IoT and AI are transforming cars with intelligent mobility. (2017, February). Retrieved from https://blogs.microsoft.com/iot/2017/02/09/how-iot-and-ai-are-transforming-cars-with-intelligent-mobility/

InfoBionic receives FDA 510(k) clearance for MoMe® Kardia. (2016, May). Retrieved from http://www.prnewswire.com/news-releases/infobionic-receives-fda-510k-clearance-for-mome-kardia-300236811.html

Lathan, C. (2016, June). 3 ways AI and robotics will transform healthcare. Retrieved from https://www.weforum.org/agenda/2016/06/3-ways-ai-and-robotics-will-transform-healthcare/

Martin, L. (2017, March). OpenAPS offers open source tools for diabetes management. Retrieved from https://makezine.com/2017/03/31/openaps/

McGrath, J. (2017, July). Tech is making life in Barcelona better, even if you don't know it's there. Retrieved from https://www.digitaltrends.com/home/barcelona-smart-city-technology/

OneTouch Vibe™ Plus insulin pump earns FDA approval and Health Canada License and is first pump integrated with the Dexcom G5® Mobile Continuous Glucose Monitor. (2016, December). Retrieved from http://www.prnewswire.com/news-releases/one touch-vibe-plus-insulin-pump-earns-fda-approval-and-health-canada-license-and-is-first-pump-integrated-with-the-dexcom-g5-mobile-continuous-glucose-monitor-300381728.html

Peters, A. (2015, May 4). This smart plate tells you how many calories you're eating–and whether you're eating too fast. Retrieved from https://www.fastcompany.com/3045810/this-smart-plate-tells-you-how-many-calories-youre-eating-and-whether-y

Rosenblum, A. (2015, December). This couple is hacking the insulin pump. Retrieved from http://www.popsci.com/hacker-medicine

Solar HUB open collaboration. (n.d.). Retrieved from http://www.rethinkthecity.com/

The internet of things and connected cars. (2017). Retrieved from http://assets1.dxc.tech nology/analytics/downloads/DXC-Analytics-the_Internet_of_Things_and_connected _cars-4AA6-5105ENW.pdf

The Smart City sector takes root in Catalonia. (n.d.). Retrieved from http://catalonia.com
/en/trade-with-catalonia/smart-city.jsp

Trafton, A. (2015, November). A new way to monitor vital signs. Retrieved from http://
news.mit.edu/2015/ingestible-sensor-measures-heart-breathing-rates-1118

Trafton, A. (2017, February). Engineers harness stomach acid to power tiny sensors. Retrieved
from http://news.mit.edu/2017/engineers-harness-stomach-acid-power-tiny-sensors
-0206

Smart Retail

Dacko, S. G. (2017, November). Enabling smart retail settings via mobile augmented reality
shopping apps. *Technological Forecasting and Social Change. Technological Forecasting
and Social Change 124*, 243–256.

Di Rienzo, A., Garzotto, F., Cremonesi, P., Frà, C., & Valla, M. (2015, September). Towards
a smart retail environment. In *Adjunct Proceedings of the 2015 ACM International
Joint Conference on Pervasive and Ubiquitous Computing and Proceedings of the 2015
ACM International Symposium on Wearable Computers* (pp. 779–782). ACM.

García, M. R., Cabo, M. L., Herrera, J. R., Ramilo-Fernández, G., Alonso, A. A., & Balsa-
Canto, E. (2017). Smart sensor to predict retail fresh fish quality under ice storage.
Journal of Food Engineering, 197, 87–97.

Hwangbo, H., Kim, Y. S., & Cha, K. J. (2017). Use of the smart store for persuasive marketing and immersive customer experiences: A case study of Korean apparel enterprise. Mobile Information Systems, 2017.

Javornik, A., Rogers, Y., Moutinho, A. M., & Freeman, R. (2016). Revealing the shop-
per experience of using a" magic mirror" augmented reality make-up application. In
Conference on Designing Interactive Systems (Vol. 2016, pp. 871–882). Association for
Computing Machinery (ACM).

Kahl, G., Klusch, M., Zinnikus, I., Schimmelpfennig, J., & Zapp, M. (2015, December).
ADIGE: Semantic business process management for smart retail environments. In
*Proceedings of the 17th International Conference on Information Integration and Web-
based Applications & Services* (p. 53). ACM.

Kim, H. Y., Lee, J. Y., Mun, J. M., & Johnson, K. K. (2017). Consumer adoption of smart
in-store technology: Assessing the predictive value of attitude versus beliefs in the
technology acceptance model. *International Journal of Fashion Design, Technology and
Education, 10*(1), 26–36.

Nowodzinski, P., Łukasik, K., & Puto, A. (2016). Internet of things (IoT) in a retail environment. The new strategy for firm's development. *European Scientific Journal, 12*(10).

Novotny, Á., Dávid, L., & Csáfor, H. (2015). Applying RFID technology in the retail
industry-benefits and concerns from the consumer's perspective. *Amfiteatru Economic,
17*(39), 615.

Priporas, C. V., Stylos, N., & Fotiadis, A. K. (2017). Generation Z consumers' expectations
of interactions in smart retailing: A future agenda. *Computers in Human Behavior,
77*, 374–381

Verhoef, P. C., Stephen, A. T., Kannan, P. K., Luo, X., Abhishek, V., Andrews, M., ...
& Hu, M. (2017). Consumer connectivity in a complex, technology-enabled, and
mobile-oriented world with smart products. *Journal of Interactive Marketing, 40*,
1–8.

Wu, B. F., Tseng, W. J., Chen, Y. S., Yao, S. J., & Chang, P. J. (2016, July). An intelligent self-checkout system for smart retail. In *2016 International Conference on System Science and Engineering (ICSSE)* (pp. 1–4). IEEE.

You, T. F. (2016). Applying the internet of things (IoT) technology to develop a new business model of smart department stores. Retrieved from http://etd.lib.nsysu.edu.tw /ETD-db/ETD-search/view_etd?URN=etd-0607116-221638

Smart Home

Amadeo, M., Campolo, C., Iera, A., & Molinaro, A. (2015, June). Information centric networking in IoT scenarios: The case of a smart home. In *2015 IEEE International Conference on Communications* (pp. 648–653). IEEE.

Anvari-Moghaddam, A., Monsef, H., & Rahimi-Kian, A. (2015). Optimal smart home energy management considering energy saving and a comfortable lifestyle. *IEEE Transactions on Smart Grid, 6*(1), 324–332.

Balta-Ozkan, N., Boteler, B., & Amerighi, O. (2014). European smart home market development: Public views on technical and economic aspects across the United Kingdom, Germany and Italy. *Energy Research & Social Science, 3*, 65–77.

Fernandes, F., Morais, H., Vale, Z., & Ramos, C. (2014). Dynamic load management in a smart home to participate in demand response events. *Energy and Buildings, 82*, 592–606.

Han, J., Choi, C. S., Park, W. K., Lee, I., & Kim, S. H. (2014). Smart home energy management system including renewable energy based on ZigBee and PLC. *IEEE Transactions on Consumer Electronics, 60*(2), 198–202.

Jacobsson, A., Boldt, M., & Carlsson, B. (2016). A risk analysis of a smart home automation system. *Future Generation Computer Systems, 56*, 719–733.

Ji, J., Liu, T., Shen, C., Wu, H., Liu, W., Su, M., ... & Jia, Z. (2016, August). A human-centered smart home system with wearable-sensor behavior analysis. In *2016 IEEE International Conference on Automation Science and Engineering* (pp. 1112–1117). IEEE.

Kumar, S. (2014). Ubiquitous smart home system using android application. *International Journal of Computer Networks and Communications, 6*(1), 33–43.

Le, T., Reeder, B., Chung, J., Thompson, H., & Demiris, G. (2014). Design of smart home sensor visualizations for older adults. *Technology and Health Care, 22*(4), 657–666.

Li, M., & Lin, H. J. (2015). Design and implementation of smart home control systems based on wireless sensor networks and power line communications. *IEEE Transactions on Industrial Electronics, 62*(7), 4430–4442.

Mennicken, S., Vermeulen, J., & Huang, E. M. (2014, September). From today's augmented houses to tomorrow's smart homes: New directions for home automation research. In *Proceedings of the 2014 ACM International Joint Conference on Pervasive and Ubiquitous Computing* (pp. 105–115). ACM.

Mowad, M. A. E. L., Fathy, A., & Hafez, A. (2014). Smart home automated control system using android application and microcontroller. *International Journal of Scientific & Engineering Research, 5*(5), 935–939.

Nawaz, A., Helbostad, J. L., Skjæret, N., Vereijken, B., Bourke, A., Dahl, Y., & Mellone, S. (2014, June). Designing smart home technology for fall prevention in older people. In *International Conference on Human-Computer Interaction* (pp. 485–490). Springer, Cham.

Santoso, F. K., & Vun, N. C. (2015, June). Securing IoT for smart home system. In *2015 IEEE International Symposium on Consumer Electronics* (pp. 1–2). IEEE.

Strengers, Y., & Nicholls, L. (2017). Convenience and energy consumption in the smart home of the future: Industry visions from Australia and beyond. *Energy Research & Social Science, 32*, 86–93.

Wan, J., O'Grady, M. J., & O'Hare, G. M. (2015). Dynamic sensor event segmentation for real-time activity recognition in a smart home context. *Personal and Ubiquitous Computing, 19*(2), 287–301.

Xu, K., Wang, X., Wei, W., Song, H., & Mao, B. (2016). Toward software defined smart home. *IEEE Communications Magazine, 54*(5), 116–122.

Zhou, S., Wu, Z., Li, J., & Zhang, X. P. (2014). Real-time energy control approach for smart home energy management system. *Electric Power Components and Systems, 42*(3–4), 315–326.

Zhou, B., Li, W., Chan, K. W., Cao, Y., Kuang, Y., Liu, X., & Wang, X. (2016). Smart home energy management systems: Concept, configurations, and scheduling strategies. *Renewable and Sustainable Energy Reviews, 61*, 30–40.

Smart Energy

Aktas, A., Erhan, K., Ozdemir, S., & Ozdemir, E. (2017). Experimental investigation of a new smart energy management algorithm for a hybrid energy storage system in smart grid applications. *Electric Power Systems Research, 144*, 185–196.

Calvillo, C. F., Sánchez-Miralles, A., & Villar, J. (2016). Energy management and planning in smart cities. *Renewable and Sustainable Energy Reviews, 55*, 273–287.

Costanza, E., Fischer, J. E., Colley, J. A., Rodden, T., Ramchurn, S. D., & Jennings, N. R. (2014, April). Doing the laundry with agents: A field trial of a future smart energy system in the home. In *Proceedings of the SIGCHI Conference on Human Factors in Computing Systems* (pp. 813–822). ACM.

El-Hawary, M. E. (2014). The smart grid—State-of-the-art and future trends. *Electric Power Components and Systems, 42*(3–4), 239–250.

Fadaeenejad, M., Saberian, A. M., Fadaee, M., Radzi, M. A. M., Hizam, H., & AbKadir, M. Z. A. (2014). The present and future of smart power grid in developing countries. *Renewable and Sustainable Energy Reviews, 29*, 828–834.

Guo, Z., Wang, Z. J., & Kashani, A. (2015). Home appliance load modeling from aggregated smart meter data. *IEEE Transactions on Power Systems, 30*(1), 254–262.

Lund, H., Werner, S., Wiltshire, R., Svendsen, S., Thorsen, J. E., Hvelplund, F., & Mathiesen, B. V. (2014). 4th generation district heating (4GDH): Integrating smart thermal grids into future sustainable energy systems. *Energy, 68*, 1–11.

Lund, P. D., Mikkola, J., & Ypyä, J. (2015). Smart energy system design for large clean power schemes in urban areas. *Journal of Cleaner Production, 103*, 437–445.

Mathiesen, B. V., Lund, H., Connolly, D., Wenzel, H., Østergaard, P. A., Möller, B., ... & Hvelplund, F. K. (2015). Smart energy systems for coherent 100% renewable energy and transport solutions. *Applied Energy, 145*, 139–154.

Naus, J., Spaargaren, G., van Vliet, B. J., & van der Horst, H. M. (2014). Smart grids, information flows and emerging domestic energy practices. *Energy Policy, 68*, 436–446.

Pan, J., Jain, R., Paul, S., Vu, T., Saifullah, A., & Sha, M. (2015). An internet of things framework for smart energy in buildings: Designs, prototype, and experiments. *IEEE Internet of Things Journal, 2*(6), 527–537.

Quilumba, F. L., Lee, W. J., Huang, H., Wang, D. Y., & Szabados, R. L. (2015). Using smart meter data to improve the accuracy of intraday load forecasting considering customer behavior similarities. *IEEE Transactions on Smart Grid, 6*(2), 911–918.

Schultz, P. W., Estrada, M., Schmitt, J., Sokoloski, R., & Silva-Send, N. (2015). Using in-home displays to provide smart meter feedback about household electricity consumption: A randomized control trial comparing kilowatts, cost, and social norms. *Energy, 90,* 351–358.

Siano, P. (2014). Demand response and smart grids—A survey. *Renewable and Sustainable Energy Reviews, 30,* 461–478.

Toft, M. B., Schuitema, G., & Thøgersen, J. (2014). Responsible technology acceptance: Model development and application to consumer acceptance of smart grid technology. *Applied Energy, 134,* 392–400.

Yang, L., Entchev, E., Rosato, A., & Sibilio, S. (2017). Smart thermal grid with integration of distributed and centralized solar energy systems. *Energy, 122,* 471–481.

Zhou, K., Fu, C., & Yang, S. (2016). Big data driven smart energy management: From big data to big insights. *Renewable and Sustainable Energy Reviews, 56,* 215–225.

Smart Security

Abas, K., Porto, C., & Obraczka, K. (2014). Wireless smart camera networks for the surveillance of public spaces. *Computer, 47*(5), 37–44.

Baranwal, T., & Pateriya, P. K. (2016, January). Development of IoT based smart security and monitoring devices for agriculture. In *2016 6th International Conference, Cloud System and Big Data Engineering* (pp. 597–602). IEEE.

Chilipirea, C., Ursache, A., Popa, D. O., & Pop, F. (2016, September). Energy efficiency and robustness for IoT: Building a smart home security system. In the *2016 IEEE 12th International Conference on Intelligent Computer Communication and Processing* (pp. 43–48). IEEE.

Chitnis, S., Deshpande, N., & Shaligram, A. (2016). An investigative study for smart home security: Issues, challenges and countermeasures. *Wireless Sensor Network, 8,* 61–68.

Elmaghraby, A. S., & Losavio, M. M. (2014). Cyber security challenges in smart cities: Safety, security and privacy. *Journal of Advanced Research, 5*(4), 491–497.

Fernández-Ares, A. J., Mora-Garcia, A. M., García-Arenas, M. I., García-Sánchez, P., Romero, G., Odeh, S. M., & Castillo, P. A. (2016). A novel wireless mobility monitoring and tracking system: Applications for smart traffic. *International Journal of Conceptual Structures and Smart Applications (IJCSSA), 4*(2), 55–71.

García, C. G., Meana-Llorián, D., G-Bustelo, B. C. P., Lovelle, J. M. C., & Garcia-Fernandez, N. (2017). Midgar: Detection of people through computer vision in the internet of things scenarios to improve the security in smart cities, smart towns, and smart homes. *Future Generation Computer Systems, 76,* 301–313.

Kumar, S., & Lee, S. R. (2014, June). Android based smart home system with control via Bluetooth and internet connectivity. In *18th IEEE International Symposium on Consumer Electronics (ISCE 2014)* (pp. 1–2). IEEE.

Lapray, P. J., Heyrman, B., & Ginhac, D. (2016). HDR-ARtiSt: An adaptive real-time smart camera for high dynamic range imaging. *Journal of Real-Time Image Processing, 12*(4), 747–762.

Lee, H. W., Liu, C. H., Chu, K. T., Mai, Y. C., Hsieh, P. C., Hsu, K. C., & Tseng, H. C. (2015). Kinect who's coming—Applying Kinect to human body height measurement to improve character recognition performance. *Smart Science, 3*(2), 117–121.

Meinel, L., Findeisen, M., Hes, M., Apitzsch, A., & Hirtz, G. (2014, January). Automated real-time surveillance for ambient assisted living using an omnidirectional camera. In *2014 IEEE International Conference on Consumer Electronics* (pp. 396–399). IEEE.

Mowad, M. A. E. L., Fathy, A., & Hafez, A. (2014). Smart home automated control system using android application and microcontroller. *International Journal of Scientific & Engineering Research, 5*(5), 935–939.

Prasad, S., Mahalakshmi, P., Sunder, A. J. C., & Swathi, R. (2014). Smart surveillance monitoring system using raspberry PI and PIR sensor. *International Journal of Computer Science and Information Technologies, 5*(6), 7107–7109.

Sahani, M., Nanda, C., Sahu, A. K., & Pattnaik, B. (2015, March). Web-based online embedded door access control and home security system based on face recognition. In *2015 International Conference on Circuit, Power and Computing Technologies* (pp. 1–6). IEEE.

Yang, S. (2014). Design and implementation of smart camera network for efficient wide area surveillance. *International Journal of Energy, Information, and Communications, 5*(1), 1–8.

Smart Automotive/Connected Car

Alheeti, K. M. A., Gruebler, A., & McDonald-Maier, K. D. (2015, September). An intrusion detection system against black hole attacks on the communication network of self-driving cars. In *2015 Sixth International Conference on Emerging Security Technologies* (pp. 86–91). IEEE.

Fagnant, D. J., & Kockelman, K. (2015). Preparing a nation for autonomous vehicles: Opportunities, barriers and policy recommendations. *Transportation Research Part A: Policy and Practice, 77*, 167–181.

Gerla, M., Lee, E. K., Pau, G., & Lee, U. (2014, March). Internet of vehicles: From intelligent grid to autonomous cars and vehicular clouds. In *2014 IEEE World Forum on Internet of Things* (pp. 241–246). IEEE.

Golestan, K., Soua, R., Karray, F., & Kamel, M. S. (2016). Situation awareness within the context of connected cars: A comprehensive review and recent trends. *Information Fusion, 29*, 68–83.

Guillet, A., Lenain, R., Thuilot, B., & Martinet, P. (2014). Adaptable robot formation control: Adaptive and predictive formation control of autonomous vehicles. *IEEE Robotics & Automation Magazine, 21*(1), 28–39.

Howard, D., & Dai, D. (2014). Public perceptions of self-driving cars: The case of Berkeley, California. In *Transportation Research Board 93rd Annual Meeting, 14*(4502).

Kang, S., Han, S., Cho, S., Jang, D., Choi, H., & Choi, J. W. (2016). High speed CAN transmission scheme supporting data rate of over 100 Mb/s. *IEEE Communications Magazine, 54*(6), 128–135.

Kirk, R. (2015). Cars of the future: The internet of things in the automotive industry. *Network Security, 2015*(9), 16–18.

Lin, P. (2016). Why ethics matters for autonomous cars. In *Autonomous driving* (pp. 69–85). Springer Berlin Heidelberg.

Mahmood, A., Casetti, C., Chiasserini, C. F., Giaccone, P., & Harri, J. (2016, January). Mobility-aware edge caching for connected cars. In *2016 12th Annual Conference on Wireless On-demand Network Systems and Services* (pp. 1–8). IEEE.

Menze, M., & Geiger, A. (2015). Object scene flow for autonomous vehicles. In *Proceedings of the IEEE Conference on Computer Vision and Pattern Recognition* (pp. 3061–3070).

Svahn, F., Lindgren, R., & Mathiassen, L. (2015, January). Applying options thinking to shape generativity in digital innovation: An action research into connected cars. In *2015 48th Hawaii International Conference on System Sciences* (pp. 4141–4150). IEEE.

van den Berg, V. A., & Verhoef, E. T. (2016). Autonomous cars and dynamic bottleneck congestion: The effects on capacity, value of time and preference heterogeneity. *Transportation Research Part B: Methodological, 94,* 43–60.

Wagner, M., & Koopman, P. (2015). A philosophy for developing trust in self-driving cars. In *Road Vehicle Automation 2* (pp. 163–171). Springer, Cham.

Smart Healthcare

Appelboom, G., Camacho, E., Abraham, M. E., Bruce, S. S., Dumont, E. L., Zacharia, B. E., ... & Connolly, E. S. (2014). Smart wearable body sensors for patient self-assessment and monitoring. *Archives of Public Health, 72*(1), 28.

Catarinucci, L., De Donno, D., Mainetti, L., Palano, L., Patrono, L., Stefanizzi, M. L., & Tarricone, L. (2015). An IoT-aware architecture for smart healthcare systems. *IEEE Internet of Things Journal, 2*(6), 515–526.

Chen, M., Ma, Y., Song, J., Lai, C. F., & Hu, B. (2016). Smart clothing: Connecting human with clouds and big data for sustainable health monitoring. *Mobile Networks and Applications, 21*(5), 825–845.

Chiuchisan, I., Costin, H. N., & Geman, O. (2014, October). Adopting the internet of things technologies in health care systems. In *2014 International Conference and Exposition on Electrical and Power Engineering* (pp. 532–535). IEEE.

Hossain, M. S., & Muhammad, G. (2016). Cloud-assisted industrial internet of things (iiot)–enabled framework for health monitoring. *Computer Networks, 101,* 192–202.

Lim, S., & Cha, M. (2017). An analysis of customized personal health information for smart healthcare systems. *Indian Journal of Forensic Medicine & Toxicology, 11*(2), 452–456.

Mandl, K. D., Kohane, I. S., McFadden, D., Weber, G. M., Natter, M., Mandel, J., ... & Adams, W. G. (2014). Scalable collaborative infrastructure for a learning health-care system (SCILHS): Architecture. *Journal of the American Medical Informatics Association, 21*(4), 615–620.

Moosavi, S. R., Gia, T. N., Rahmani, A. M., Nigussie, E., Virtanen, S., Isoaho, J., & Tenhunen, H. (2015). SEA: a secure and efficient authentication and authorization architecture for IoT-based healthcare using smart gateways. *Procedia Computer Science, 52,* 452–459.

Muhammad, G. (2015). Automatic speech recognition using interlaced derivative pattern for cloud based healthcare system. *Cluster Computing, 18*(2), 795–802.

Porter, M. E., & Heppelmann, J. E. (2014). How smart, connected products are transforming competition. *Harvard Business Review, 92*(11), 64–88.

Rahmani, A. M., Thanigaivelan, N. K., Gia, T. N., Granados, J., Negash, B., Liljeberg, P., & Tenhunen, H. (2015, January). Smart e-health gateway: Bringing intelligence to internet-of-things based ubiquitous healthcare systems. In *2015 12th Annual IEEE Consumer Communications and Networking Conference* (pp. 826–834). IEEE.

Sakr, S., & Elgammal, A. (2016). Towards a comprehensive data analytics framework for smart healthcare services. *Big Data Research, 4*, 44–58.

Solanas, A., Patsakis, C., Conti, M., Vlachos, I. S., Ramos, V., Falcone, F., ... & Martinez-Balleste, A. (2014). Smart health: A context-aware health paradigm within smart cities. *IEEE Communications Magazine, 52*(8), 74–81.

Stantchev, V., Barnawi, A., Ghulam, S., Schubert, J., & Tamm, G. (2015). Smart items, fog and cloud computing as enablers of servitization in healthcare. *Sensors & Transducers, 185*(2), 121.

Tarapiah, S., Aziz, K., Atalla, S., & Ismail, S. H. (2016). Smart real-time healthcare monitoring and tracking system using GSM/GPS technologies. *International Journal of Computer Applications, 142*, 19–26.

Vargiu, E., & Zambonelli, F. (2017). Engineering IoT systems through agent abstractions: Smart healthcare as a case study. In *Agents and multi-agent systems for health care* (pp. 25–39). Springer, Cham.

Vashist, S. K., Schneider, E. M., & Luong, J. H. (2014). Commercial smartphone-based devices and smart applications for personalized healthcare monitoring and management. *Diagnostics, 4*(3), 104–128.

Yang, L., Ge, Y., Li, W., Rao, W., & Shen, W. (2014, May). A home mobile healthcare system for wheelchair users. In *Proceedings of the 2014 IEEE 18th International Conference on Computer Supported Cooperative Work in Design* (pp. 609–614). IEEE.

Yu, R., Mak, T. W., Zhang, R., Wong, S. H., Zheng, Y., Lau, J. Y., & Poon, C. C. (2017, May). Smart healthcare: Cloud-enabled body sensor networks. In *2017 IEEE 14th International Conference on Wearable and Implantable Body Sensor Networks* (pp. 99–102). IEEE.

Yuehong, Y. I. N., Zeng, Y., Chen, X., & Fan, Y. (2016). The internet of things in healthcare: An overview. *Journal of Industrial Information Integration, 1*, 3–13.

Chapter 5

IoT-Enabled Smart Cities

I have an affection for a great city. I feel safe in the neighborhood of man, and enjoy the sweet security of the streets.

Henry Wadsworth Longfellow

After reading this chapter you will be able to:

- Understand the concept of smart cities
- Understand the potential uses of IoT in smart cities like Barcelona and Singapore
- Know the relevance of smart IT infrastructure for smart cities
- Gain insight about the necessity of smart, inclusive cities
- Know the key performance indicators for smart, sustainable cities
- Interpret the City Resilience Framework
- Understand how data can enable smart cities

Introduction

With an ever-increasing population in cities worldwide, it has become a humongous task to provide basic as well as enhanced facilities to city populations spanning all spheres of life. The number of urban residents is growing by approximately 60 million every year, and by 2050 it is estimated that people occupying just two percent of the world's land will consume about three-quarters of its resources.* To provide a better and connected experience, the concept of smart cities is being explored in various parts of the globe as a viable alternative to the current state of metropolitan life.

---

* Mitchell, S. et al. (2013). The Internet of everything for cities. Retrieved from http://www .cisco.com/web/strategy/docs/gov/everything-for-cities.pdf

Information and communication technology (ICT) has enhanced business across the globe, first as an enabler and then as an integrated function of business. Now the smart city concept aims to enhance human experience in cities in the near future through ICT-enabled and IoT-embedded services. Also, smart cities are being projected as information-economy hubs for creating a knowledge society. And to make a city smart, there should be a risk-managed, secured underlying digital infrastructure that provides real-time data for all services and assisted living in the city.

The Concept of the Smart City

The British Standards Institution (BSI) defines a smart city as a "city that effectively integrates the physical, spatial, digital and human worlds to deliver a sustainable, prosperous and inclusive future for its citizens."*

As shown in Figure 5.1, one of the key features of a smart city is that it has a citizen-centric approach with a digitally enabled infrastructure. The U.K. Department of Business Innovation and Skills has laid out specific criteria for smart cities:

> A Smart City should enable every citizen to engage with all the services on offer, public as well as private, in a way best suited to his or her needs. It brings together hard infrastructure, social capital including local skills and community institutions, and (digital) technologies to fuel sustainable economic development and provide an attractive environment for all.†

Smart-assisted living in smart cities includes features like smart infrastructure, smart energy management, smart transport and traffic management, smart water management, smart waste management, smart healthcare and smart education. The term 'smart' mostly implies an ICT-enabled or embedded intelligence based on the real-time experience of services.

According to the IEEE:

A smart city brings together technology, government and society to enable the following characteristics:

- A smart economy
- Smart mobility
- A smart environment
- Smart people
- Smart living
- Smart governance."‡

---

* BSI. (2014). PAS 181 smart city framework. Retrieved from http://www.bsigroup.com/en-GB /smart-cities/Smart-Cities-Standards-and-Publication/PAS-181-smart-cities-framework/
† Department for Business Innovation & Skills, UK Government. (2013). Smart cities: Background paper. https://www.gov.uk/government/uploads/system/uploads/attachment _data/file/246019/bis-13-1209-smart-cities-background-paper-digital.pdf
‡ IEEE (2015). Smart cities. Retrieved from http://smartcities.ieee.org/about.html

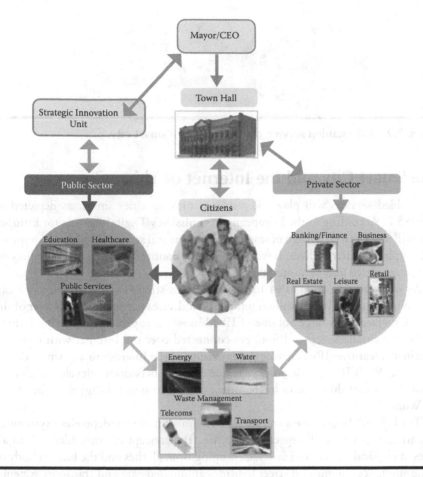

Figure 5.1   The smart city model.

Department of Business Innovation & Skills, UK Government. (2013). Smart Cities: Background Paper. Retrieved from https://www.gov.uk/government/uploads /system/uploads/attachment_data/file/246019/bis-13-1209-smart-cities-background -paper-digital.pdf.

While various services are being redesigned to make existing cities smarter, there are plans for new smart cities to be created in the next two decades across the globe to meet the growing needs of the human population. For example, cities like San Jose, Singapore and Barcelona are implementing smart services for their citizens, discussed in detail in this chapter, while India has ambitious plans to develop 100 smart cities* across the country in a decade.

---

* Tolan C. (2014, July) Cities of the future? Indian PM pushes plan for 100 'smart cities.' *CNN.com*. Retrieved from http://edition.cnn.com/2014/07/18/world/asia/india-modi-smart -cities/index.html

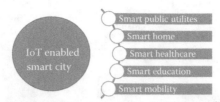

Figure 5.2 IoT-enabled service components of a smart city.

The Smart City and the Internet of Things

IoT-enabled services will play a key role in making cities smart, as depicted in Figure 5.2. According to the European Commission's 'Digital Agenda for Europe,' "IoT will enable an ecosystem of smart applications and services, which will improve and simplify"[*] citizens' lives. It will bring hyper-connectivity to a global society by using augmented and rich interfaces.

As we transition from IPv4 to the already existing IPv6 technology, we can have more than two billion billion unique IP addresses per square millimeter of the Earth's surface. This huge volume of IP addresses is capable of digitizing human civilization in all aspects of life to get connected over the Internet with uniquely identifiable features. IPv6 enables the extension of the Internet to any smart device or service. With IPv6, "it is possible to use a global network to develop one's own network of smart things or to interconnect one's own smart things with the rest of the World."[†]

The IoT will help create a connected environment of interdependent systems in a smart city, covering all aspects of city life. The concept of embedded IoT in all types of physical objects and artifacts, ranging from clothes and the human body to home appliances, home and street lighting, automated cars and transport systems, and public utilities will transform cities into smart cities.

The IoT-enabled live digital fabric of interdependent systems in a smart city will be dynamic in nature, with instantaneous data gathering and analytics features. The synergistic interdependency of systems in a smart city will provide immense opportunities for enhanced living quality and governance. It can help smart-city councils strategize necessary actions and governance based on continuous analytics of the huge volume of data collected from subsystems at every moment. Data collected from IoT-enabled smart services can be utilized for managing the energy efficiency of buildings, mapping social data for crime prevention, monitoring flood situations, public consultations and trend analysis, and city development in areas like housing, education, transport, medical services and employment. Another

---

[*] European Commission. (2015). The Internet of things. Retrieved from http://ec.europa.eu /digital-agenda/en/internet-things

[†] IPv6 for IoT. (n.d.). Retrieved from http://iot6.eu/ipv6_for_iot

potential opportunity is the decreasing cost of service operations through enhanced efficiency and reduced wastage of natural resources.

Interdependent Systems and Smart IT Infrastructure for Smart Cities

Smart cities' interdependent systems will provide critical infrastructure and smart services to handle all major public systems and citizen services. These include water and energy generation and transmission, logistics, transport, waste disposal, green buildings, street and home lighting, connected healthcare for citizens, online learning and education, surveillance and traffic. The interdependent systems in a smart city will enable dynamic, synergistic data gathering and analytics. In effect, this creates a 'system of systems' that generally follows a scale-free topology to allow expansion without affecting the attributes of interdependency and interconnectedness.

The Japan Electronics and Information Technology Industries Association (JEITA) has created a social-infrastructure model: JEITA I-model 2.0.* It includes five layers of IT infrastructure that can be utilized in designing the information architecture of smart cities. These include:

a. **Hardware**. This layer is a component of the social infrastructure and subsystems like intelligent transportation systems and power generation systems. It consists of hardware and embedded software such as sensors, actuators and information terminals.

b. **Embedded software**. This is the system layer playing the role of supplying information obtained from devices and subsystems, such as sensors and actuators, to the upper layer.

c. **Databank**. This Infrastructure layer is utilized for data publishing and maintenance of data acquired by the embedded software layer.

d. **Platform**. This is the foundational information system for providing safety, security and ease-of-use features required by different applications for exchange of data between various services with high reliability and control.

e. **Application**. This layer shows the various applications in the smart social infrastructure. It has a data-processing function of each smart service, a data acquisition function that automatically acquires the information scattered over the web, an ability to integrate various types of data obtained from a plurality of infrastructures and an ability to manage the collected data with features such as prediction and optimization, visualization and real-time processing capabilities.

---

* JEITA (2013). JEITA I-model 2.0. Retrieved from http://home.jeita.or.jp/cgi-bin/page/detail .cgi?n=581&ca=1

Key Performance Indicators for Smart Sustainable Cities

The International Telecommunication Union's Focus Group on Smart Sustainable Cities has defined a smart sustainable city as "an innovative city that uses information and communication technologies (ICTs) and other means to improve quality of life, efficiency of urban operation and services, and competitiveness, while ensuring that it meets the needs of present and future generations with respect to economic, social and environmental aspects."*

The Focus Group has described four main aspects of smart sustainable cities. These are:

a. The **economic** ability to generate income and employment for the livelihood of the inhabitants
b. The **social** ability to ensure the well-being of citizens in the domains of safety, health and education, irrespective of the citizens' class, race or gender
c. The **environmental** ability to protect natural resources for future quality and reproducibility
d. The ability to **govern** the social conditions for stability, democracy, participation and justice

The Focus Group has also identified key performance indicators (KPIs) to evaluate ICT's contribution to making cities smarter and sustainable. The KPIs have been categorized into the following six dimensions, shown in Figure 5.3:

1. Information and communication technology
2. Environmental sustainability
3. Productivity
4. Quality of life
5. Equity and social inclusion
6. Physical infrastructure

The sub-dimensions for each of the above dimensions have been identified by the Focus Group as shown in Table 5.1 below.

Smart, Inclusive Cities for All—A Global Vision

The Global Initiative for Inclusive ICTs (G3ict), an advocacy initiative launched by the United Nations Global Alliance for ICT and Development, has partnered with

---

* International Telecommunications Union. (2014). ITU-T FG-SSC: ITU-T focus group on smart sustainable cities: Overview of key performance indicators in smart sustainable cities. Retrieved from https://www.itu.int/en/ITU-T/focusgroups/ssc/Documents/Approved _Deliverables/TS-Overview-KPI.docx

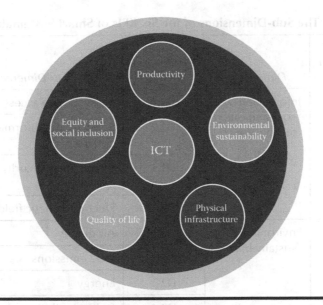

Figure 5.3   The six dimensions of KPIs for smart sustainable cities.

International Telecommunications Union. (2014). ITU-T FG-SSC: ITU-T focus group on smart sustainable cities: Overview of key performance indicators in smart sustainable cities. Retrieved from https://www.itu.int/en/ITU-T/focusgroups/ssc /Documents/Approved_Deliverables/TS-Overview-KPI.docx

World Enabled, a global education, communications and strategic consulting group, to launch the Smart Cities for All (SC4A) global initiative to develop a new tool to assess and benchmark digital inclusion and accessibility in smart cities through the SC4A Digital Inclusion Maturity Model.* This maturity model will quantify the progress of smart cities toward achieving ICT accessibility and digital inclusion across a broad range of functions for older persons and persons with disabilities.

As per the SC4A initiative[†] launched in June 2016:

■ The importance of digital technologies is increasing and smart cities are booming, with at least 88 smart cities expected to be operational globally by 2025.
■ There is a growing digital divide for older persons with persons with disabilities.

---

* Smart Cities for All. (2017, October). Press Release: Smart Cities for All Digital Inclusion Maturity Model launch. Retrieved from http://smartcities4all.org/20171030_Press_Release _WCD_Draft_XT_v12.php
† Smart Cities for All. (n.d.). Communicating the Case for a Stronger Commitment to Digital Inclusion in Cities. Retrieved from http://smartcities4all.org/SC4A_Communicating_the _Case_XT.php

Table 5.1   The Sub-Dimensions of the Six KPIs of Smart Sustainable Cities*

Dimension Number	Dimension	Sub-Dimension Number	Sub-Dimension
D1	Information and communication technology	D1.1	Network and access
		D1.2	Services and information platforms
		D1.3	Information security and privacy
		D1.4	Electromagnetic field
D2	Environmental sustainability	D2.1	Air quality
		D2.2	CO2 emissions
		D2.3	Energy
		D2.4	Indoor pollution
		D2.5	Water, soil and noise
D3	Productivity	D3.1	Capital investment
		D3.2	Employment
		D3.3	Inflation
		D3.4	Trade
		D3.5	Savings
		D3.6	Export/import
		D3.7	Household income/ consumption
		D3.8	Innovation
		D3.9	Knowledge economy
D4	Quality of life	D4.1	Education
		D4.2	Health
		D4.3	Safety/security public place
		D4.4	Convenience and comfort

(Continued)

Table 5.1 (Continued)   The Sub-Dimensions of the Six KPIs of Smart Sustainable Cities*

Dimension Number	Dimension	Sub-Dimension Number	Sub-Dimension
D5	Equity and social inclusion	D5.1	Inequity of income/consumption (Gini coefficient)
		D5.2	Social and gender inequity of access to services and infrastructure
		D5.3	Openness and public participation
		D5.4	Governance
D6	Physical infrastructure	D6.1	Infrastructure/connection to services—piped water
		D6.2	Infrastructure/connection to services—sewage
		D6.3	Infrastructure/connection to services—electricity
		D6.4	Infrastructure/connection to services—waste management
		D6.5	Connection to services—knowledge infrastructure
		D6.6	Infrastructure/connection to services—health infrastructure
		D6.7	Infrastructure/connection to services—transport
		D6.8	Infrastructure/connection to services—road infrastructure
		D6.9	Housing—building materials
		D6.10	Housing—living space
		D6.11	Building

*International Telecommunications Union. (2014). ITU-T FG-SSC: ITU-T focus group on smart sustainable cities: Overview of key performance indicators in smart sustainable cities. Retrieved from https://www.itu.int/en/ITU-T/focusgroups/ssc/Documents/Approved_Deliverables/TS-Overview-KPI.docx

- While our reliance on digital devices continues to grow, persons with disabilities are being largely excluded from digital accessibility even though they make up a large population in cities, and with their close friends and relatives, they have a disposable income of over $8 trillion globally.
- Smart cities should have a strong focus on ICT accessibility, and policies should align smart-city programs with ICT accessibility objectives and assistive-technology standards to make future smart cities inclusive for all.

Smart cities are now coming up with innovative solutions to become more inclusive for all citizens. For example, Singapore, with its 'Smart Nation' vision, has developed a prototype of a four-wheel self-driving personal mobility device* as a collaborative effort from the Singapore-MIT Alliance for Research and Technology. This device will provide mobility to persons with reduced mobility and complement the existing transportation system.

The City Resilience Framework

As shown in Figure 5.4, the City Resilience Framework,[†] developed by Arup International Development with support from the Rockefeller Foundation, determines the drivers that contribute to the resilience of cities. With the help of this framework, cities can assess the extent of their resilience and identify critical areas of weakness and the relevant actions and programs to make improvements.

City resilience has been defined as "the capacity of cities to function so that people living in cities . . . survive and thrive no matter what shocks and stresses they encounter."[‡] As mentioned below, the City Resilience Framework is comprised of four dimensions, 12 goals and seven basic qualities:

1. The health and well-being of everyone (**people**) living and working in the city.
2. An economy and society (**organization**) that enable the urban population to live peacefully and act collectively.
3. Urban systems and services (**place**) that empower stakeholders and integrated planning
4. Leadership and strategy (**knowledge**) comprising of man-made and natural infrastructures that provide critical services and protects urban citizens

---

* Smart Nation and Digital Government Office. (n.d.). Self-Driving Vehicles (SDVs): Future of Mobility in Singapore. Retrieved from https://www.smartnation.sg/initiatives/Mobility/self-driving-vehicles-sdvs--future-of-mobility-in-singapore
† The Rockefeller Foundation. (2014). City resilience framework. Retrieved from https://assets.rockefellerfoundation.org/app/uploads/20140410162455/City-Resilience-Framework-2015.pdf
‡ Ibid.

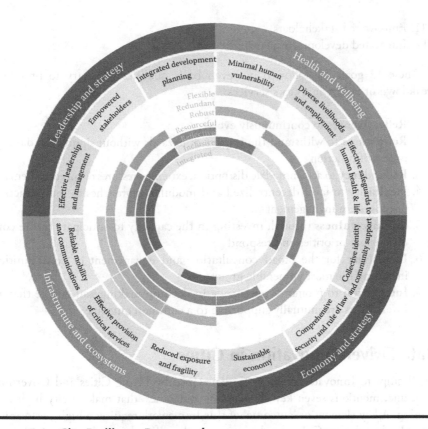

Figure 5.4   City Resilience Framework.

The Rockefeller Foundation. (2014). City resilience framework. Retrieved from https://assets.rockefellerfoundation.org/app/uploads/20140410162455/City -Resilience-Framework-2015.pdf

The 12 goals relate to the elements of a city's immune system. Compromising any of these elements will require compensating with strength in the other areas, or else the overall resilience of the city gets affected. The 12 goals of a resilient city should be:

1. Minimal human vulnerability
2. Diverse livelihood and employment
3. Effective safeguards to human health and life
4. Collective identity and community support
5. Comprehensive security and rule of law
6. Sustainable economy
7. Reduced exposure and fragility (infrastructure and environment)
8. Effective provision of critical services
9. Reliable communication and mobility
10. Effective leadership and management

11. Empowered stakeholders
12. Integrated development planning

These 12 goals are followed by seven basic qualities necessary to prevent a breakdown or failure in a city's services:

1. **Reflectiveness** to continuously evolve
2. **Robustness** to withstand the impact of hazards without significant damage or loss of function
3. **Redundancy** to accommodate disruption, extreme pressures or surges in demand
4. **Flexibility** to favor decentralized and modular approaches to infrastructure or ecosystem management
5. **Resourcefulness** through investing in the capacity to anticipate future conditions, set priorities and respond
6. **Inclusivity** for the broad consultation and engagement of communities, including the most vulnerable groups
7. **Integration** to promote consistency in decision making and ensure that all investments are mutually supportive to a common outcome

Data-Driven Innovation in Cities

The 'Equipt to Innovate'* framework, developed by Living Cities and Governing Magazine, mentions seven key elements or 'outcomes' that make a city high performing. A key element or 'outcome' of this framework requires a high-performing city to be data driven. Cities should appropriately utilize data and modern technologies for better performance, innovation and engagement. Data should be regularly captured on the city's progress toward its desired results, and these data should be regularly reviewed. Government and vendor data should be publicly available and easy to understand by the citizens for public review and analysis. Data analytics should be applied by the government to proactively support better outcomes for city services. Use of free, lightweight and open-source technological tools should be encouraged by the cities along with enterprise software solutions to allow an effective flow of information between city systems and stakeholders for innovative activities.

How Cities Are Becoming Smart

In this section we will discuss various initiatives of the world's leading cities that make these cities truly smart, inclusive and innovative. Some significant smart-city projects in Barcelona and Singapore have been highlighted here. Various other initiatives are also being executed across the globe to make cities ready for smart living.

---

* Field guide—Equip to innovate. (n.d.). Retrieved from http://www.governing.com/equipt
  /about-the-equipt-framework.html#About+the+Equipt+Framework

Figure 5.5   Smart city: Barcelona.

Sheu, T. (2015, April). Barcelona's Quest to Become The Smartest City of Them All. Retrieved from http://www.barcinno.com/smart-city-barcelona/

Smart City: Barcelona

As shown in Figure 5.5, Barcelona's smart-city initiative aims to provide better and affordable citizen services through digital transformation in order to make the government more transparent, participative and effective. As a part of this initiative, high-speed internet connectivity is considered an absolute necessity, and not a luxury. The key areas of smart-city planning for Barcelona are digital transformation, digital innovation and digital empowerment. Cisco began experimenting with turning Barcelona into a smart city in 2011.

The three key visions for digital transformation in Barcelona* are:

1. Technology for better government
2. Urban technology
3. City data commons

For the 'technology for better government'[†] vision, the city of Barcelona's government has detected social problems, prioritized them and planned strategic projects to address these problems. Some of the projects to realize this vision are discussed below:

▪ **Open budget**. This is a tool that provides citizens with the city council's budgetary data of the current financial year and those of previous years. Users can know where public money goes by browsing, analyzing and understanding budgetary allocations to the highest level of detail, and can make

---

* Barcelona Digital City. (n.d.). Digital transformation. Retrieved from http://ajuntament .barcelona.cat/digital/en/digital-transformation

† Barcelona Digital City. (n.d.). Technology for a better government. Retrieved from http://ajun tament.barcelona.cat/digital/en/digital-transformation/technology-for-a-better-government

comparisons between the amounts budgeted and spent, referring to invoices of actual budgetary expenditures.

■ **Digital market.** This is a relationship between the city council and technology companies to speed up public-procurement processes with transparency in public-contracting processes and accountability. This digital environment enhances collaboration between companies, shortens the contracting process, lowers the access barrier to public contracts for small- and medium-sized businesses and also facilitates innovative public-procurement processes.

■ **Ethical mailbox.** The Office for Transparency and Good Practices (OTBP) of the city of Barcelona has provided a digital channel named the 'ethical mailbox' for the general public to report corruption. It also enables the general public and municipal personnel to report unlawful and unethical conduct with confidentiality, anonymity and indemnity. Civil society organizations are also collaborating in this initiative to combat corruption.

Sensors have been connected to trash cans to request to be emptied by garbage trucks only when they are full. Smart-parking initiative include app-based facilities that help to identify vacant parking slots and also help drivers pay on-street parking fees by using mobile devices instead of having to use parking meters. The city's lamp posts have turned digital, with individual IP addresses, and can monitor crowds, traffic and local weather so police are well informed of accidents as they happen in real time. Barcelona city officials also plan on creating the first city-based operating system, CityOS, to run the city from a single interface.

A key initiative to enable *'urban technology'** in Barcelona is Sentilo. This is an open-source sensor and actuator platform to exploit the information generated by the city of Barcelona. As shown in the figure below, a layer of sensors has been deployed across the city to collect and broadcast city information. This model of architecture is now being used by other cities in their digital-smartness journeys.

Sentilo enables the compilation and sharing of data from IoT devices, as depicted in Figure 5.6. Barcelona aims to connect Sentilo to CityOS to offer data to the general public through the OpenData portal. Sentilo is also being offered as a 'software as a service' through Thingtia, a Sentilo-based, IoT-multipurpose system built by Opentrends, which is one of the sponsors of Sentilo.

CityOS is a significant digital transformation initiative as part of forming a *'city data commons'.*[†] This is a city-based operating system that aims to use data generated in the city to offer better services and to optimize Barcelona's internal processes. The city council of Barcelona is developing the CityOS architecture based on Big Data technology, with features of better data governance, quality controls,

---

* Barcelona Digital City. (n.d.). Sentilo. Retrieved from http://ajuntament.barcelona.cat /digital/en/digital-transformation/urban-technology/sentilo
† Barcelona Digital City. (n.d.). CityOS. Retrieved from http://ajuntament.barcelona.cat /digital/en/digital-transformation/city-data-commons/cityos

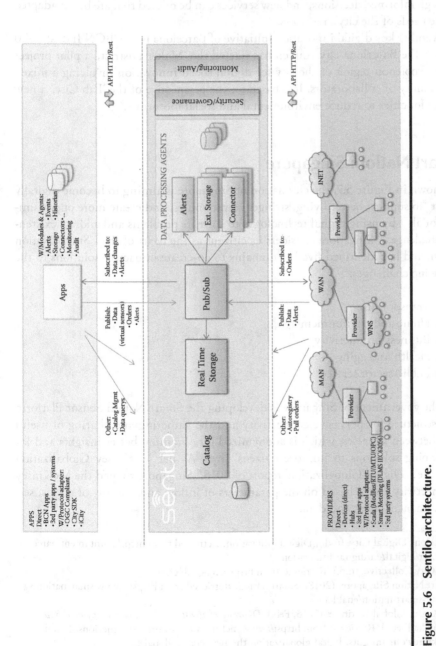

Figure 5.6 Sentilo architecture.

Sentilo. (2013). Architecture. Retrieved from http://www.sentilo.io/xwiki/bin/view/Sentilo.Community.Documentation/Architecture

effective processing and security. CityOS will provide a unified point of access to data from the internal processes of the city council and from external agencies. It will help to provide structured decision making based on data from various city council areas. As a result, municipal resources can be distributed in a better way through informed decisions, and new services can be offered that are better adapted to the needs of the city's residents.

Another key digital innovation initiative of Barcelona is the 'BCN Industry 4.0 Hub.'* The Barcelona city council has started the 'Maker District,' a pilot project in the Poblenou region of the city for digital social innovation involving a mixed community of collaborators. It is based on the perspective of the 'Fab City,' a new model for cities to reduce environmental and social footprints.[†]

Smart Nation: Singapore

As shown in Figure 5.7, the city-nation of Singapore is aiming to become digitally smart "to support better living, stringer communities, and create more opportunities for all"[‡] by using digital technology "to solve its problems and address existential challenges." Singapore considers its citizens as the heart of their Smart Nation vision, and has identified five key domains to co-create impactful solutions to the following challenges:

1. Transportation
2. Home and environment
3. Business productivity
4. Health and aging
5. Public-sector services

The government of Singapore is developing the Smart Nation Sensor Platform infrastructure for pervasive connectivity and the gathering and sharing of useful data between agencies, which is anonymized for obtaining better insights and for developing solutions to improve citizens' lives. As per McKinsey Globalization Institute's 'Digital Globalization' report of 2016,[§] Singapore topped the McKinsey Connectivity Index based on the parameters of inflow and outflow of goods, services, finance, talent and data.

---

* Barcelona Digital City. (n.d.). Digital innovation. Retrieved from http://ajuntament.barce lona.cat/digital/en/digital-innovation
† Fab City Collective. (n.d.). Retrieved from https://blog.fab.city/
‡ Smart Nation Singapore. (2018). Smart nation. Retrieved from https://www.smartnation.sg /about-smart-nation/enablers
§ McKinsey Global Institute. (2016, Feb.). *Digital globalization: The new era of global flows.* Manyika J. et al. Retrieved from https://www.mckinsey.com/business-functions/digital -mckinsey/our-insights/digital-globalization-the-new-era-of-global-flows

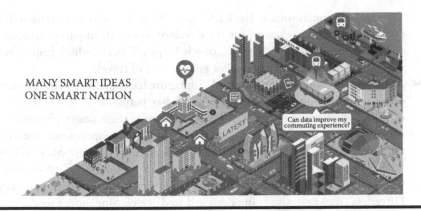

Figure 5.7   Smart nation: Singapore.

Smart Nation Singapore. (2018). Many smart ideas one smart nation. Retrieved from https://www.smartnation.sg/

Some of the digital initiatives to make Singapore a smart nation* are discussed below.

Mobility

The challenge: scarcity of land, a growing population and more than a million vehicles in Singapore require an optimized use of the limited road and transportation infrastructure for more efficient, safe, reliable and enhanced transportation.

Some digital opportunities identified for smart mobility in Singapore are:

■ **Contactless fare payment for public transportation.** Singapore has a vision of a 'hands-free' fare system utilizing multiple options like cards, mobile devices and wearables. An account-based ticketing system has been designed by the Land Transport Authority (LTA) and MasterCard for direct payments for public bus and train rides with contactless credit and debit cards. There are initiatives to utilize health and fitness wearable devices as contactless payment modes for public transport rides as well. Near field communication-enabled phones are being used to tap in and out of mass rapid transit (MRT) and light rail transit (LRT) public buses and taxis.

■ **Open data analysis for urban transportation.** Singapore's LTA is digitally capturing the arrival times of public buses by tracking sensors installed on these vehicles for real-time location analysis and planning of fleets to meet surges in demand for commuters. This initiative reduced crowding issues in bus services by 92%, and also reduced average waiting time by up to seven

---

* Smart Nation Singapore. (2018). Initiatives. Retrieved from https://www.smartnation.sg /initiatives

minutes on popular routes. The LTA has publicly shared transportation data, and is creating an ecosystem for the development of third-party transportation applications. One such application is Happy Wheels, which helps wheelchair users find accessible routes for greater ease of travel.

- **Autonomous mobility on demand.** Singapore has launched trials for autonomous mobility-on-demand services, and has plans for shared self-driving shuttles for commuters that can be booked from their smart phones. Self-driving vehicles have been performing road tests in Singapore since as early as 2015. The National University of Singapore and the Singapore-MIT Alliance for Research and Technology have developed a prototype for a four-wheel, self-driving 'personal mobility device' for persons with reduced mobility.* Driverless trucks are also being designed to drive on Singapore's roads.

Living

The challenge: a safe, sustainable and livable urban environment in spite of land constraints.

Some digital opportunities for smart living[†] in Singapore are:

- **The One Service mobile application.** The Municipal Services Office and the Natural Environment Agency of Singapore have developed the One Service mobile application to obtain citizens' feedback, which is routed to the appropriate agencies for necessary action.
- **Smart-home solutions.** Smart-home solutions are being tried in Singapore's homes to evaluate and identify the right smart applications to help residents achieve greater convenience, reduce utility expenses and provide elderly care.

Health

The challenge: technology-assisted healthcare initiatives to aid active aging for an increased number of elderly citizens.

Some digital opportunities for smart health[‡] in Singapore are:

- **Assisted Robotics.** Assisted robotics have been applied to develop RoboCoach, which ensures that elderly persons perform their routine exercises correctly through motion-sensing technology. RoboCoach also provides physical and

---

* Smart Nation Singapore. (2018). Self-driving vehicles (SDVs): Future of mobility in Singapore. Retrieved from https://www.smartnation.sg/initiatives/Mobility/self-driving-vehicles-sdvs--future-of-mobility-in-singapore
† Smart Nation Singapore. (2018). Living. Retrieved from https://www.smartnation.sg/initiatives/Living/
‡ Smart Nation Singapore. (2018). Health. Retrieved from https://www.smartnation.sg/initiatives/Health/

cognitive therapy to elderly people who suffer strokes or have disorders like Alzheimer's or Parkinson's.

■ **Health Hub**. The Health Hub web portal and mobile application has been launched as Singapore's first one-stop online health information and service portal. It helps users access hospital records, lab test reports and medical records and make medical appointments. The 'Caregiver Access' module allows patients to grant caregivers access to their personal health records, medical records and medical appointment information.

■ **The Telehealth initiative**. Integrated and seamless healthcare services at home have been started with the TeleHealth initiative. This aims to improve access to rehabilitation services for better functional recovery of patients. For patients with orthopedic conditions, data is transmitted wirelessly from wearable motion sensors to clinicians and therapists to help them remotely conduct rehabilitation and therapy sessions. A video-conferencing function is also supported for one-to-one or multi-user interactions and consultations.

Services

The challenge: serving the needs of citizens and empower communities to seek help and to help one another.

Some digital opportunities for smart services* in Singapore are:

■ **The My Info platform for residents**. My Info, a consent-based platform for Singapore residents, helps residents manage their personal data for e-services of the government. Over 150 government digital services can be conveniently accessed from this platform. CorpPlan has been created for authentication and authorization and for businesses and other entities to access the government's e-services. The online portal eCitizen has been provided for citizens to access a host of government services and e-services.

■ **Contactless payment**. Singapore aims to become a smart, cashless society with contactless payment. Widespread acceptance of contactless payment is being encouraged by the government. Consumers from participating banks in Singapore can securely transact in peer-to-peer mode using the recipient's mobile number and/or National Registration Identity Card number.

■ **A Fintech regulatory sandbox**. Singapore has also proposed creating a regulatory sandbox for Fintech experimentation to identify means for the financial industry to increase its operational efficiency, create economic opportunities and better manage risks.

---

* Smart Nation Singapore. (2018). Services. Retrieved from https://www.smartnation.sg /initiatives/Services/

Suggested Reading

Albino, V., Berardi, U., & Dangelico, R. M. (2015). Smart cities: Definitions, dimensions, performance, and initiatives. *Journal of Urban Technology, 22*(1), 3–21.

Al Nuaimi, E., Al Neyadi, H., Mohamed, N., & Al-Jaroodi, J. (2015). Applications of big data to smart cities. *Journal of Internet Services and Applications, 6*(1), 25.

Angelidou, M. (2016). Four European smart city strategies. *International Journal of Social Science Studies, 4*(4), 18–30.

Angelidou, M. (2015). Smart cities: A conjuncture of four forces. *Cities, 47*, 95–106.

Angelidou, M. (2014). Smart city policies: A spatial approach. *Cities, 41*, S3–S11.

Anthopoulos, L. G., Janssen, M., & Weerakkody, V. (2015, May). Comparing smart cities with different modeling approaches. In *Proceedings of the 24th International Conference on World Wide Web* (pp. 525–528). ACM.

Anttiroiko, A. V., Valkama, P., & Bailey, S. J. (2014). Smart cities in the new service economy: Building platforms for smart services. *AI & Society, 29*(3), 323–334.

Boulos, M. N. K., Tsouros, A. D., & Holopainen, A. (2015). 'Social, innovative and smart cities are happy and resilient': Insights from the WHO EURO 2014 International Healthy Cities Conference. *International Journal of Health Geographics, 14*(1), 3.

Boulos, M. N. K., & Al-Shorbaji, N. M. (2014). On the internet of things, smart cities and the WHO Healthy Cities. *International Journal of Health Geographics, 13*(1), 10.

Cagliero, L., Cerquitelli, T., Chiusano, S., Garino, P., Nardone, M., Pralio, B., & Venturini, L. (2015, April). Monitoring the citizens' perception on urban security in smart city environments. In *2015 31st IEEE International Conference on Data Engineering Workshop* (pp. 112–116). IEEE.

Carvalho, L. (2014). Smart cities from scratch? A socio-technical perspective. *Cambridge Journal of Regions, Economy and Society, 8*(1), 43–60.

Clohessy, T., Acton, T., & Morgan, L. (2014, December). Smart City as a Service (SCaaS): A future roadmap for e-government smart city cloud computing initiatives. In *Proceedings of the 2014 IEEE/ACM 7th International Conference on Utility and Cloud Computing* (pp. 836–841). IEEE Computer Society.

Cocchia, A. (2014). Smart and digital city: A systematic literature review. In *Smart city* (pp. 13–43). Springer International Publishing.

Dameri, R. P., & Rosenthal-Sabroux, C. (2014). Smart city and value creation. In *Smart city* (pp. 1–12). Springer International Publishing.

Delmastro, F., Arnaboldi, V., & Conti, M. (2016). People-centric computing and communications in smart cities. *IEEE Communications Magazine, 54*(7), 122–128.

Djahel, S., Doolan, R., Muntean, G. M., & Murphy, J. (2015). A communications-oriented perspective on traffic management systems for smart cities: Challenges and innovative approaches. *IEEE Communications Surveys & Tutorials, 17*(1), 125–151.

Elmaghraby, A. S., & Losavio, M. M. (2014). Cyber security challenges in smart cities: Safety, security and privacy. *Journal of Advanced Research, 5*(4), 491–497.

Franke, T., Lukowicz, P., & Blanke, U. (2015). Smart crowds in smart cities: Real life, city scale deployments of a smartphone based participatory crowd management platform. *Journal of Internet Services and Applications, 6*(1), 27.

Gabrys, J. (2014). Programming environments: Environmentality and citizen sensing in the smart city. *Environment and Planning D: Society and Space, 32*(1), 30–48.

Glasmeier, A., & Christopherson, S. (2015). Thinking about smart cities. *Cambridge Journal of Regions, Economy and Society, 8*, 3–12.

Imteaj, A., Chowdhury, M., & Mahamud, M. A. (2015, May). Dissipation of waste using dynamic perception and alarming system: A smart city application. In the *2015 International Conference on Electrical Engineering and Information Communication Technology (ICEEICT)*, (pp. 1–5). IEEE.

Jensen, O. B. (2016). Drone city-power, design and aerial mobility in the age of "smart cities". *Geographica Helvetica, 71*(2), 67.

Jin, J., Gubbi, J., Marusic, S., & Palaniswami, M. (2014). An information framework for creating a smart city through internet of things. *IEEE Internet of Things Journal, 1*(2), 112–121.

Khan, Z., Anjum, A., Soomro, K., & Tahir, M. A. (2015). Towards cloud based big data analytics for smart future cities. *Journal of Cloud Computing, 4*(1), 2.

Kitchin, R. (2014). The real-time city? Big data and smart urbanism. *GeoJournal, 79*(1), 1–14.

Kitchin, R. (2015). Making sense of smart cities: Addressing present shortcomings. *Cambridge Journal of Regions, Economy and Society, 8*(1), 131–136.

Kitchin, R. (2016). The ethics of smart cities and urban science. *Philosophical Transactions of the Royal Society*. R. Soc. A, 74(2083), 20160115.

Kylili, A., & Fokaides, P. A. (2015). European smart cities: The role of zero energy buildings. *Sustainable Cities and Society, 15*, 86–95.

Lau, S. P., Merrett, G. V., Weddell, A. S., & White, N. M. (2015). A traffic-aware street lighting scheme for Smart Cities using autonomous networked sensors. *Computers & Electrical Engineering, 45*, 192–207.

Lee, J., & Lee, H. (2014). Developing and validating a citizen-centric typology for smart city services. *Government Information Quarterly, 31*, S93–S105.

Letaifa, S. B. (2015). How to strategize smart cities: Revealing the SMART model. *Journal of Business Research, 68*(7), 1414–1419.

Ma, M., Preum, S. M., Tarneberg, W., Ahmed, M., Ruiters, M., & Stankovic, J. (2016, May). Detection of runtime conflicts among services in smart cities. In the *2016 IEEE International Conference on Smart Computing* (pp. 1–10). IEEE.

Marsal-Llacuna, M. L., Colomer-Llinàs, J., & Meléndez-Frigola, J. (2015). Lessons in urban monitoring taken from sustainable and livable cities to better address the smart cities initiative. *Technological Forecasting and Social Change, 90*, 611–622.

Mattoni, B., Gugliermetti, F., & Bisegna, F. (2015). A multilevel method to assess and design the renovation and integration of smart cities. *Sustainable Cities and Society, 15*, 105–119.

Medvedev, A., Zaslavsky, A., Khoruzhnikov, S., & Grudinin, V. (2015). Reporting road problems in smart cities using open IoT framework. In *Interoperability and open-source solutions for the internet of things* (pp. 169–182). Springer International Publishing.

Medvedev, A., Fedchenkov, P., Zaslavsky, A., Anagnostopoulos, T., & Khoruzhnikov, S. (2015, August). Waste management as an IoT-enabled service in smart cities. In *Conference on Smart Spaces* (pp. 104–115). Springer International Publishing.

Mohammed, F., Idries, A., Mohamed, N., Al-Jaroodi, J., & Jawhar, I. (2014, May). UAVs for smart cities: Opportunities and challenges. *In 2014 International Conference on Unmanned Aircraft Systems* (pp. 267–273). IEEE.

Ojo, A., Curry, E., & Janowski, T. (2014). Designing next generation smart city initiatives-harnessing findings and lessons from a study of ten smart city programs. In *Proceedings on the European Conference on Information Systems ECIS 2014* (pp. 1–15). Tel Aviv, Israel.

Perera, C., Zaslavsky, A., Christen, P., & Georgakopoulos, D. (2014). Sensing as a service model for smart cities supported by internet of things. *Transactions on Emerging Telecommunications Technologies, 25*(1), 81–93.

Piro, G., Cianci, I., Grieco, L. A., Boggia, G., & Camarda, P. (2014). Information centric services in smart cities. *Journal of Systems and Software, 88*, 169–188.

Puiu, D., Barnaghi, P., Toenjes, R., Kümper, D., Ali, M. I., Mileo, A., ... & Gao, F. (2016). Citypulse: Large scale data analytics framework for smart cities. *IEEE Access, 4*, 1086–1108.

Ramaswami, A., Russell, A. G., Culligan, P. J., Sharma, K. R., & Kumar, E. (2016). Meta-principles for developing smart, sustainable, and healthy cities. *Science, 352*(6288), 940–943.

Robert, J., Kubler, S., Kolbe, N., Cerioni, A., Gastaud, E., & Främling, K. (2017). Open IoT ecosystem for enhanced interoperability in smart cities—Example of Métropole De Lyon. *Sensors, 17*(12), 2849.

Sanchez, L., Muñoz, L., Galache, J. A., Sotres, P., Santana, J. R., Gutierrez, V., ... & Pfisterer, D. (2014). SmartSantander: IoT experimentation over a smart city testbed. *Computer Networks, 61*, 217–238.

Sen, J., Majumdar, M., Saha, D., & Chaudhuri, A. (2017). Smart economy in smart cities Varanasi India: Case of a smart traditional economy of knowledge-based institutional services and creative-cultural products. In *Smart economy in smart cities* (pp. 553–576). Springer, Singapore.

Tei, K., & Gurgen, L. (2014, March). ClouT: Cloud of things for empowering the citizen clout in smart cities. In the *2014 IEEE World Forum on Internet of Things* (pp. 369–370). IEEE.

Vanolo, A. (2014). Smartmentality: The smart city as disciplinary strategy. *Urban Studies, 51*(5), 883–898.

Walravens, N. (2015). Qualitative indicators for smart city business models: The case of mobile services and applications. *Telecommunications Policy, 39*(3), 218–240.

Zanella, A., Bui, N., Castellani, A., Vangelista, L., & Zorzi, M. (2014). Internet of things for smart cities. *IEEE Internet of Things Journal, 1*(1), 22–32.

'BY THINGS'

Chapter 6

IoT Security and Privacy Concerns

Liberty requires security without intrusion, security plus privacy.

Bruce Schneier

After reading this chapter you will be able to:

- Understand the security concerns due to constraints of IoT devices
- Understand the security and privacy concerns of IoT data, networks and applications
- Gain an insight on scenario-based security and privacy concerns of smart healthcare, smart billing and payment services, smart homes, smart retail, proximity marketing and smart vending machines.

Introduction

IoT-enabled smart services are not yet fully secured. In addition, there are privacy concerns for IoT-enabled service offerings that deal with user data, user-owned device data and data from the environment encapsulating the user or IoT device.

Security and privacy concerns are major roadblocks to the successful deployment of IoT-enabled smart services. The "TCS Global Trend study on IoT"* has found that ensuring security and reliability is one of the three biggest factors holding companies back from realizing the promise of IoT for businesses across all sectors.

Based on 396 responses to a sample survey limited to enterprises that expected that IoT would impact their business, Gartner[†] reported in 2015 that more than 50% of responding enterprises mentioned security and privacy concerns as the top inhibitors to the adoption of the IoT.

The scale and complexity of IoT-enabled services make it challenging to implement traditional security techniques for the IoT. This is because the IoT presents a unique set of access control challenges due to the low power requirements of IoT devices, the inability of these devices to run complex encryption algorithms due to memory limitations and the distributed nature of the extremely large number of IoT devices required to create a system of systems for providing context-aware services.

According to NIST, a variety of security concerns arise from the constraints of power consumption, price factor and lifecycle variation in IoT devices, as shown in Figure 6.1[‡] below.

Adversaries can physically or remotely intercept and manipulate data captured by IoT sensor nodes. Data transmission from these sensors and gateway devices can be passively monitored if encryption is not robust. Malicious nodes can be embedded in wireless sensor networks to access data and communicate with neighbor nodes.

Privacy is another critical concern for IoT-enabled services, because personally identifiable information can be gathered from gateway devices without consent for business benefits and unscrupulous activities.

Security and privacy of IoT-enabled smart services acquire a bigger dimension than traditional IT-enabled services due to the pervasive nature of these devices and services. There are also other technical and policy-related challenges of the IoT that need to be addressed, as shown in Figure 6.2. Here, we discuss some possible scenarios to understand the security and privacy concerns that can come up in IoT-enabled services.

---

* TCS Global Trend Study. (2015, July). Internet of things: the complete reimaginative force. Retrieved from http://sites.tcs.com/internet-of-things/wp-content/uploads/Internet-of-Things -The-Complete-Reimaginative-Force.pdf
[†] Gartner Inc. 2015. Mass adoption of the internet of things will create new opportunities and challenges for enterprises. Retrieved from http://www.gartner.com/technology/research.jsp
[‡] IoT cybersecurity considerations. (2017, June). Retrieved from https://www.nist.gov/itl /applied-cybersecurity/iot-cybersecurity-considerations#technical

Consideration	Device constraint	Security concern
Power consumption	Many IoT devices require a long battery life, without access to a permanent power supply.	Power-efficient hardware may lack additional capabilities like ability to support encryption or hardware security mechanisms.
Low cost	The consumer's perceived value of a device greatly depends oon the cost to purchase and implement the device. Market drivers often require that companies produce devices at a very low cost.	In meeting these price pressures, devices may have low processing power and constrained hardware space, offering limited support for security mechanisms.
Lifecycle	The lifespan of devices vary greatly, some devices (like simple sensors) are short-lived, while others are meant to last for decades.	Over time, devices may become hardware-constrained and cannot be updated. Built in security mechanisms may be found vulnerable or deprecated, like old encryption suites.

Figure 6.1   Security concerns due to constraints of IoT devices.

IoT cybersecurity considerations. (2017, June). Retrieved from https://www.nist .gov/itl/applied-cybersecurity/iot-cybersecurity-considerations#technical

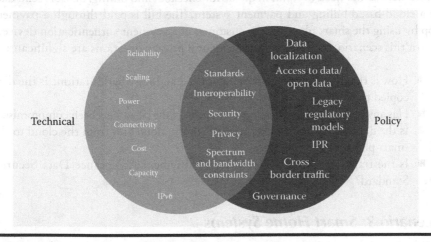

Figure 6.2   The IoT's emerging challenges.

Biggs, P., Garrity, J., LaSalle, C., & Polomska, A. (n.d.). Harnessing the internet of things for global development. Retrieved from https://www.sbs.ox.ac .uk/cybersecurity-capacity/system/files/Harnessing-IoT-Global-Development.pdf

Scenario 1: Smart Healthcare Systems

IoT-enabled smart healthcare services in a smart city can provide opportunities to remotely monitor the health of diabetics and patients with kidney and heart problems through data gathered from pacemakers and smart glucose monitors implanted in the patients' bodies and exchanged on a 24/7 basis directly to citizens' health monitoring systems in smart hospitals. In such a critical service, if access to a glucose monitor or pacemaker is breached, or the authenticity of the data obtained from medical devices cannot be verified, then the patients' life might be in risk.

Security and privacy concerns in this scenario are:

■ Who has access to the private medical details of patients?
■ Is the data sent from sensors to the gateway device encrypted?
■ Is the data stored at the gateway device?
■ How much personally identifiable information about the patient is being captured and stored?
■ Is the personally identifiable information anonymized?
■ How to verify what information is sent back to the wearable medical device from the remote monitoring center?

Scenario 2: Smart Billing and Payment Systems

In a retail outlet, IoT sensors can be used in carts to add up purchases by customers. Customers do not need to stand in queue for checkout and billing. Sensors send data to a cloud-based billing and payment system. The bill is paid through a payment app by using the smart phone of the customer as a payment authentication device.

In this scenario, the following security and privacy questions are significant:

■ How is the data from sensors being logged and for what duration? Is the data copied to multiple locations for backup?
■ Has any personally identifiable information of the customers been compromised?
■ Is the data safe in transit from sensors to the cloud and from the cloud to the smart phone of the customer?
■ Is the transaction compliant with PCI Payment Acceptance Data Security Standard?

Scenario 3: Smart Home Systems

IoT-enabled home security and temperature control systems use sensors to collect data from multiple edge components and share that data with various mobile devices. If an attacker can gain access to these smart systems through malicious means, then he can modify the functional logic of the home security and temperature control systems, resulting in a compromise of the physical security of the residents.

The security and privacy concerns in this scenario are:

■ What data is captured and transmitted by the IoT devices used for this service?
■ Who can access the data generated from a home security system?
■ Is the data sent to the actuator encrypted?
■ Is there any authentication of who sends data to the actuator?
■ Does the IoT product vendor have access to the data generated from these devices?

Scenario 4: Smart Fitting Rooms and Smart Dressing Areas in Retail Outlets

In a retail outlet, RFID sensors can be used in smart fitting rooms to allow customers to flip through a catalogue on a touch screen and indicate which items to display in the dressing room.* When the shopper walks into the dressing area, a smart mirror recognizes the items and displays the different clothing on its screen. Shopping behavior data of customers can be stored by the retailer for cross-selling recommendations.

In this scenario, the following security and privacy questions are significant:

■ What data is gathered and sent by the sensors?
■ Can the supply chain data be compromised during transit?
■ Does the personal data of customers that are collected by the sensors remain anonymous?
■ Is there any interception of the data gathered by the sensors?

Scenario 5: Proximity Marketing

With the IoT, we are realizing the benefits of proximity marketing by using Bluetooth-enabled beacons that can identify loyal customers in the proximity of billboards fitted with beacons with IoT sensors.† By activating an app on the customer's smart phone, customer data is gathered by a sensor and sent to the cloud for analytical processing and personalized marketing content is sent back to the customer's smartphone. For example, Apple is using iBeacons‡ and Facebook§ is offering place tips to users who rely on beacons at various locations in New York City.

In this scenario, the following security and privacy questions are significant:

■ Can beacon communication be compromised during transit?
■ Does beacon communication happen with the consent of customers?
■ Does the personal data collected by sensors remain anonymous?
■ Who can access customer data in the cloud?

---

* CSA. (2015). Security guidance for early adopters of the internet of things (IoT). Retrieved from https://downloads.cloudsecurityalliance.org/whitepapers/Security_Guidance_for_Early _Adopters_of_the_Internet_of_Things.pdf
† Ibid
‡ Tilley, A. (2014). Apple iBeacons find their way into McDonald's. Retrieved from http://www .forbes.com/sites/aarontilley/2014/12/18/mcdonalds-ibeacon/
§ Ingraham, N. (2015). Facebook place tips will try to put useful info about your location right into the News Feed. Retrieved from http://www.theverge.com/2015/1/29/7929351 /facebook-place-tips-pilot-launch

Scenario 6: Smart Vending Machines

Customers using smart vending machines select particular products from a display and customer details are tracked immediately from the customer's smart phone by NFC smart-phone payment support fitted to the vending machine for instant e-billing and payment. Data sent by the vending machine to the cloud is used by merchants for stock replenishment, health checks of the vending machine and for studying the popularity of products displayed.

The security and privacy concerns in this scenario are:

- Is the data sent from sensors to the gateway device encrypted?
- Is the customer's financial data exposed during payment?
- Can merchants exploit customer information for business benefit?
- Is any customer's identifiable information being stored in gateway devices or the cloud?
- Is the customer data collected at sensor nodes compromised by any means?

Relevant concerns of security and privacy are being raised by users and regulators across the globe for various other smart-service offerings that are being designed and deployed.

IoT Security Concerns

The security concerns for IoT-enabled services can be addressed from five dimensions as discussed below and depicted in Figure 6.3.

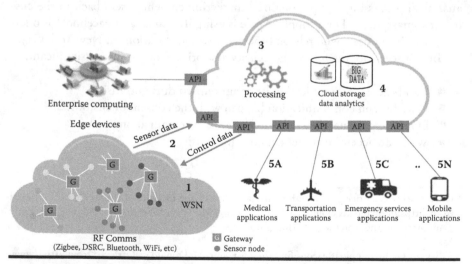

Figure 6.3   Five dimensions of security concerns for IoT services.

CSA. (2015). Security guidance for early adopters of the internet of things. Retrieved from https://downloads.cloudsecurityalliance.org/whitepapers/Security _Guidance_for_Early_Adopters_of_the_Internet_of_Things.pdf

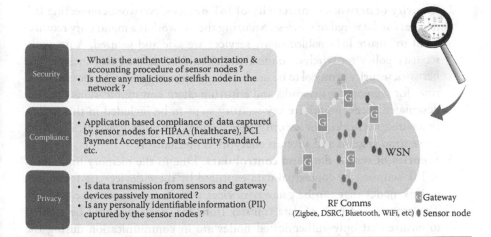

Figure 6.4   Security and privacy concerns for IoT sensors.

1. **Security of IoT sensors.** Wireless sensors provide the last mile of connectivity for IoT-enabled service implementation, and they can be damaged by human influence or by natural decay. The memory and power limitations of IoT devices also make them vulnerable to eavesdropping and radio-jamming attacks. Thus, threat mitigation in IoT devices is of utmost importance, and is the first step in ensuring a secured smart service. For smart service implementation in open, outdoor environments, periodic site surveys are required to keep track of the IoT device outlay and frequent physical checks of the devices are necessary to identify damaged sensors in need of replacement.

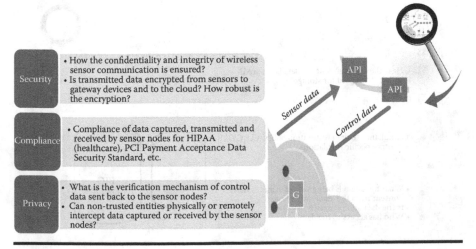

Figure 6.5   Security and privacy concerns for sensor data and control data.

2. **Security of network connectivity of IoT devices**. Network-connecting IoT devices can be wired or wireless. Securing the network is a mandatory requirement to ensure IoT-enabled smart services are safe and secured. A network-security policy for wireless connectivity should be defined that includes the network security protocol to be used. Use of a non-suggestive service set identifier for segmenting networks and ensuring client communications through designated access points are key security steps to be included in this policy. For wired networks, the security policy should include firewalls, intrusion prevention systems and a robust encryption mechanism.

3. **Security of sensor data and control data**. Due to the memory limitations of IoT sensors, the contextual data captured by these devices are forwarded to sink nodes that can be gateway devices with memory storage capacities. The process and path of data transfer from IoT devices should be secured to ensure that only authenticated nodes are in communication during the data transfer process and that data loss is prevented during the transfer. Data encryption in IoT devices using complex algorithms is not a feasible solution due to the limited computing power and energy storage capacity of these devices.

4. **Security of IoT Big Data stored locally or in the cloud**. The data collected by sensors are stored locally or sent to the cloud for analytics and application-based processing. Data-breach prevention, data-loss prevention and data-owner identification are prime considerations for the security of IoT data stored in the cloud. There can be various jurisdictional requirements regarding ownership of IoT-generated data that is copied across multiple geographies to create redundancy.

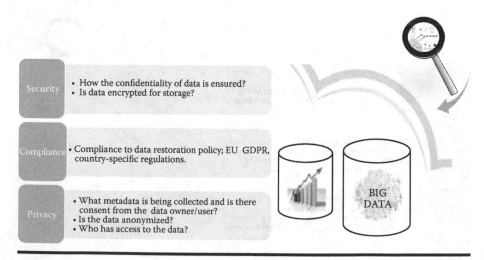

Figure 6.6   Security and privacy of IoT Big Data.

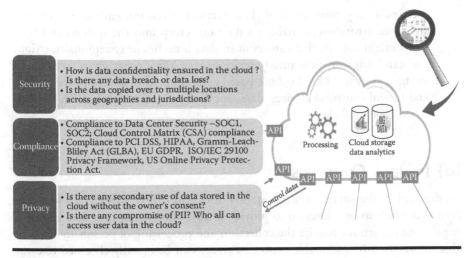

Figure 6.7   Cloud security for IoT services.

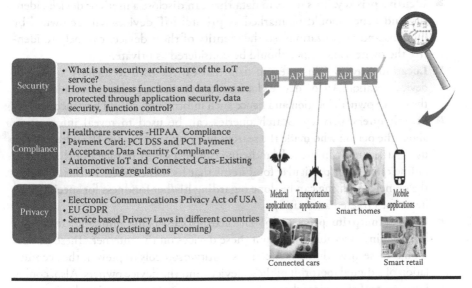

Figure 6.8   Security of IoT applications.

5. **Security for end-to-end control of devices, data, applications and networks**.
   To implement smart-service offerings, specifically in smart cities, we will have
   to interconnect huge numbers of IoT sensors, devices and gateways. This will
   create a system of systems that feed data to each other for various smart functionalities. To ensure security of these smart services we have to keep track
   and control of all devices, data, applications and networks that deliver these

services. If any component of these services is not traceable or not in control, then erroneous or malicious data can creep into the systems and incidents might happen that can result in data breaches or system malfunctions and can bring down the smart service or even affect other smart services. To mitigate security risks, attribute-based access control can be implemented for end-to-end control of devices, data, applications and networks with context-based access policies.

IoT Privacy Concerns

As depicted in the smart-service scenarios discussed earlier, privacy is a key concern that needs to be addressed to provide a trustworthy smart service. Globally, regulations on privacy require the collection and processing of personally identifiable information in a verifiable manner. Privacy can be broadly classified into the following types*:

- **Identity privacy.** This refers to data that can disclose a user's or device's identity and hence should be marked as private. IoT devices will be owned by humans and organizations, so the identity of these devices can help to identify the owners and hence should be considered as private.
- **Location privacy.** This refers to data that can be used to identify a user's or device's location. An IoT device's location can be used to gather information about the device owner's location and hence such information should be concealed.
- **Search query privacy.** Search queries can be used to reveal information about the person who made the search query by tracking the IP address of the user. An example in the IoT landscape can be a smart refrigerator that makes online queries for particular food items that its owner likes. In this scenario, the owner's profiling can be done regarding his/her fondness for specific food items for targeted advertising.
- **Digital footprint privacy.** IoT-enabled devices, being online all time, can leave behind traceable data about these devices on the Internet. These devices should be secured through effective security protocols to prevent the accumulation of a digital footprint of these devices and the device owners. Also, cookie invasion on IoT devices should be prevented to ensure operational privacy.

Protecting the metadata collected by IoT devices can be considered as a viable option to address privacy concerns. IoT metadata can have details like location,

---

* Chaudhuri, A. (2016). Addressing Security and Privacy Concerns to Realize IoT's Potential. Retrieved from https://www.tcs.com/content/dam/tcs/pdf/Services/technology-operations /Addressing%20Security%20and%20Privacy20Concerns%20to%20Realize%20IoT%27s %20Potential.pdf

FIVE QUESTIONS TO THE GLOBAL LEADER

Michelle Chibba
Instructor and strategic policy, privacy and information management advisor—
Privacy by Design Centre of Excellence.
Ryerson University
Toronto, Canada

1. **What is the significance of data privacy in realizing the potential of the Internet of Things?**

   Data privacy has everything to do with the successful uptake of the Internet of Things. We are in the early days of this emerging technology, but already we have witnessed unfortunate breaches—vulnerabilities associated with Internet-connected vehicles, residential security cameras, smart televisions, fitness wearables, to name a few. As some may say, data privacy will be the 'Achilles heel' of the Internet of Things if we don't start seriously addressing it now.

2. **Do you consider 'privacy by design' as a fundamental principle for developing trustworthy IoT products and services?**

   Privacy by Design is a global privacy framework for the 21st century. When privacy is considered early on at the conceptual stage and throughout the development lifecycle of a technology, trustworthiness is enhanced for the end user. In the IoT world, organizations must seek ways to use the wealth of knowledge they have about individuals to provide better products and services to them in ways that increase trust, not suspicion—through Privacy by Design.

3. **According to your opinion, what are the key policy features that should be considered by smart city councils to provide reliable smart services to citizens?**

   These days, "design thinking" is having a major influence on innovation, creativity and productivity. In the context of smart cities, privacy must be approached from the same design-thinking perspective and be a key consideration of smart city councils. These councils must ensure that privacy is designed into every technology, system, standard, protocol and process that touches the lives and identities of citizens in a connected community. These councils need to understand that, in the information and technology era we live in, the protection of our right to informational privacy is increasingly critical to the preservation of our rights to life, liberty, and security of the person—in essence, to preserving our freedom.

4. **What is the role of users of smart systems in ensuring the protection of their personal information?**

The best results for users of smart systems are usually those that are consciously designed around the interests and needs of individual users, who have the greatest vested interest in the management of their own personal data. The principle of user centricity emphasizes the need for human-machine interfaces to be human-centered, user-centric and user-friendly so that users may make informed privacy decisions. Information technologies, processes and infrastructures must be designed not just for individual users, but also structured by them. Users are rarely, if ever, involved in every design decision or transaction involving their personal information, but they are nonetheless in an unprecedented position today to exercise a measure of meaningful control over those designs and transactions, as well as the disposition and use of their personal information by others.

5. **How can we develop smart cognitive cities with citizen participation?**

If we are to enlist citizens in the development of innovative technologies for smart cognitive cities, then privacy and data security must be essential components in the design of such systems. The willingness to share our personal information waxes and wanes and depends upon the context in which we find ourselves. Individuals must feel comfortable that their privacy will not be violated as they move about in public spaces. It is this very willingness, based upon individual control, that lies at the heart of privacy and the choices we make about how much and to whom we wish to share the details of our lives and identities. In building cities with cognitive capabilities, the timing is ideal to make privacy the focal point for engaging citizens. Using a Privacy by Design framework will encourage innovative design approaches to be taken— ones that will enable both privacy and the desired benefits of the systems integral to smart cognitive cities.

time stamps, proximity to other IoT devices and relevant information for a device owner's profiling. The organization/city council owning the smart service should develop a cyber-privacy strategy for the IoT environment addressing the above-mentioned privacy concerns.

The IoT is a promising technological advancement that can benefit us with new digital business models and new service offerings. Businesses and city councils across the globe are willing to realize the benefits of the IoT, but they also want IoT technology to be secure and reliable. To realize the potentials of this technology we have to address the scenario-based security and privacy concerns as discussed in this chapter. A secured IoT-enabled smart service with privacy features can make it trustworthy for the ultimate benefit of users, businesses and city councils.

Suggested Reading

Alpár, G., Batina, L., Batten, L., Moonsamy, V., Krasnova, A., Guellier, A., & Natgunanathan, I. (2016, May). New directions in IoT privacy using attribute-based authentication. In *Proceedings of the ACM International Conference on Computing Frontiers* (pp. 461–466). ACM.

Apthorpe, N., Reisman, D., & Feamster, N. (2017). A smart home is no castle: Privacy vulnerabilities of encrypted IoT traffic. *arXiv preprint arXiv:1705.06805.*

Arias, O., Wurm, J., Hoang, K., & Jin, Y. (2015). Privacy and security in internet of things and wearable devices. *IEEE Transactions on Multi-Scale Computing Systems, 1*(2), 99–109.

Bertino, E., Choo, K. K. R., Georgakopolous, D., & Nepal, S. (2016). Internet of Things (IoT): Smart and secure service delivery. *ACM Transactions on Internet Technology (TOIT), 16*(4), 22.

Caron, X., Bosua, R., Maynard, S. B., & Ahmad, A. (2016). The internet of things (IoT) and its impact on individual privacy: An Australian perspective. *Computer Law & Security Review, 32*(1), 4–15.

Cirani, S., Picone, M., Gonizzi, P., Veltri, L., & Ferrari, G. (2015). IoT-oas: An oauth-based authorization service architecture for secure services in IoT scenarios. *IEEE Sensors Journal, 15*(2), 1224–1234.

Davies, N., Taft, N., Satyanarayanan, M., Clinch, S., & Amos, B. (2016, February). Privacy mediators: Helping IoT cross the chasm. In *Proceedings of the 17th International Workshop on Mobile Computing Systems and Applications* (pp. 39–44). ACM.

Farooq, M. U., Waseem, M., Khairi, A., & Mazhar, S. (2015). A critical analysis on the security concerns of internet of things (IoT). *International Journal of Computer Applications, 111*(7).

Granjal, J., Monteiro, E., & Silva, J. S. (2015). Security for the internet of things: A survey of existing protocols and open research issues. *IEEE Communications Surveys & Tutorials, 17*(3), 1294–1312.

Henze, M., Hermerschmidt, L., Kerpen, D., Häußling, R., Rumpe, B., & Wehrle, K. (2014, August). User-driven privacy enforcement for cloud-based services in the internet of things. In *International Conference on Future Internet of Things and Cloud (FiCloud 2014)* (pp. 191–196). IEEE.

Hossain, M. M., Fotouhi, M., & Hasan, R. (2015, June). Towards an analysis of security issues, challenges, and open problems in the internet of things. In *IEEE World Congress on Services (SERVICES 2015)* (pp. 21–28). IEEE.

Hwang, Y. H. (2015, April). IoT security & privacy: Threats and challenges. In *Proceedings of the 1st ACM Workshop on IoT Privacy, Trust, and Security* (pp. 1–1). ACM.

Jing, Q., Vasilakos, A. V., Wan, J., Lu, J., & Qiu, D. (2014). Security of the internet of things: Perspectives and challenges. *Wireless Networks, 20*(8), 2481–2501.

Kolias, C., Stavrou, A., Voas, J., Bojanova, I., & Kuhn, R. (2016). Learning internet-of-things security "hands-on". *IEEE Security & Privacy, 14*(1), 37–46.

Lee, J. Y., Lin, W. C., & Huang, Y. H. (2014, May). A lightweight authentication protocol for internet of things. In *International Symposium on Next-Generation Electronics (ISNE 2014)* (pp. 1–2). IEEE.

Leloglu, E. (2017). A review of security concerns in internet of things. *Journal of Computer and Communications, 5*, 121–136.

Li, S., Tryfonas, T., & Li, H. (2016). The internet of things: A security point of view. *Internet Research, 26*(2), 337–359.

Mahmoud, R., Yousuf, T., Aloul, F., & Zualkernan, I. (2015, December). Internet of things (IoT) security: Current status, challenges and prospective measures. In *10th International Conference for Internet Technology and Secured Transactions (ICITST 2015)* (pp. 336–341). IEEE.

Medaglia, C. M., & Serbanati, A. (2010). An overview of privacy and security issues in the internet of things. In *The Internet of Things* (pp. 389–395). New York: Springer.

Perera, C., Ranjan, R., Wang, L., Khan, S. U., & Zomaya, A. Y. (2015). Big data privacy in the internet of things era. *IT Professional, 17*(3), 32–39.

Pöhls, H. C., Angelakis, V., Suppan, S., Fischer, K., Oikonomou, G., Tragos, E. Z., ... & Mouroutis, T. (2014, April). RERUM: Building a reliable IoT upon privacy-and security-enabled smart objects. In *2014 Wireless Communications and Networking Conference Workshops (WCNCW)* (pp. 122–127). IEEE.

Porambage, P., Ylianttila, M., Schmitt, C., Kumar, P., Gurtov, A., & Vasilakos, A. V. (2016). The quest for privacy in the internet of things. *IEEE Cloud Computing, 3*(2), 36–45.

Riahi, A., Natalizio, E., Challal, Y., Mitton, N., & Iera, A. (2014, February). A systemic and cognitive approach for IoT security. In *2014 International Conference on Computing, Networking and Communications (ICNC)* (pp. 183–188). IEEE.

Sadeghi, A. R., Wachsmann, C., & Waidner, M. (2015, June). Security and privacy challenges in industrial internet of things. In *Design Automation Conference (DAC), 2015 52nd ACM/EDAC/IEEE* (pp. 1–6). IEEE.

Sicari, S., Rizzardi, A., Grieco, L. A., & Coen-Porisini, A. (2015). Security, privacy and trust in Internet of Things: The road ahead. *Computer Networks, 76*, 146–164.

Singh, J., Pasquier, T., Bacon, J., Ko, H., & Eyers, D. (2016). Twenty security considerations for cloud-supported internet of things. *IEEE Internet of Things Journal, 3*(3), 269–284.

Stojmenovic, I., & Wen, S. (2014, September). The fog computing paradigm: Scenarios and security issues. In *2014 Federated Conference on Computer Science and Information Systems (FedCSIS)* (pp. 1–8). IEEE.

Tankard, C. (2015). The security issues of the Internet of Things. *Computer Fraud & Security, 2015*(9), 11–14.

Thierer, A. D. (2015). The internet of things and wearable technology: Addressing privacy and security concerns without derailing innovation. *Richmond Journal of Law & Technology, 21*(2).

Trappe, W., Howard, R., & Moore, R. S. (2015). Low-energy security: Limits and opportunities in the internet of things. *IEEE Security & Privacy, 13*(1), 14–21.

Weber, R. H. (2010). Internet of Things–New security and privacy challenges. *Computer Law & Security Review, 26*(1), 23–30.

Weber, R. H. (2015). Internet of things: Privacy issues revisited. *Computer Law & Security Review, 31*(5), 618–627.

Wurm, J., Hoang, K., Arias, O., Sadeghi, A. R., & Jin, Y. (2016, January). Security analysis on consumer and industrial IoT devices. In *21st Asia and South Pacific Design Automation Conference (ASP-DAC 2016)* (pp. 519–524). IEEE.

Xu, T., Wendt, J. B., & Potkonjak, M. (2014, November). Security of IoT systems: Design challenges and opportunities. In *Proceedings of the 2014 IEEE/ACM International Conference on Computer-Aided Design* (pp. 417–423). IEEE Press.

Yi, S., Qin, Z., & Li, Q. (2015, August). Security and privacy issues of fog computing: A survey. In *International Conference on Wireless Algorithms, Systems, and Applications* (pp. 685–695). Springer, Cham.

Yoon, S., Park, H., & Yoo, H. S. (2015). Security issues on smarthome in IoT environment. In *Computer Science and its Applications* (pp. 691–696). Berlin, Heidelberg: Springer.

Yoshigoe, K., Dai, W., Abramson, M., & Jacobs, A. (2015, December). Overcoming invasion of privacy in smart home environment with synthetic packet injection. In *TRON Symposium (TRONSHOW), 2015* (pp. 1–7). IEEE.

Zhang, Z. K., Cho, M. C. Y., & Shieh, S. (2015, April). Emerging security threats and countermeasures in IoT. In *Proceedings of the 10th ACM Symposium on Information, Computer and Communications Security* (pp. 1–6). ACM.

Zhang, Z. K., Cho, M. C. Y., Wang, C. W., Hsu, C. W., Chen, C. K., & Shieh, S. (2014, November). IoT security: Ongoing challenges and research opportunities. In *2014 IEEE 7th International Conference on Service-Oriented Computing and Applications (SOCA)* (pp. 230–234). IEEE.

Zhou, J., Cao, Z., Dong, X., & Vasilakos, A. V. (2017). Security and privacy for cloud-based IoT: Challenges. *IEEE Communications Magazine, 55*(1), 26–33.

Zhou, W., & Piramuthu, S. (2014, June). Security/privacy of wearable fitness tracking IoT devices. In *2014 9th Iberian Conference on Information Systems and Technologies (CISTI),* (pp. 1–5). IEEE.

Zhou, W., & Piramuthu, S. (2015). Information relevance model of customized privacy for IoT. *Journal of Business Ethics, 131*(1), 19–30.

Chapter 7

Cyber-Threat Mitigation of Wireless Sensor Nodes for Secured and Trustworthy IoT Services

A brand is simply trust.

Steve Jobs

After reading this chapter you will be able to:

- Understand the concept of wireless sensor nodes in use for IoT applications
- Gain an insight on the security threat landscape of wireless sensor nodes
- Interpret the approach of mitigating these security threats
- Understand the assurance approach for wireless sensor security of IoT services.

Introduction

The IoT is an emerging technology that enables the interaction of uniquely identifiable computing devices that can be embedded with other interfaces like machines, linked via wired and wireless networks. An important function of the IoT is its ability to capture contextual data from the environment. The monitoring and management systems for smart services are exposed to the environment through sensors to create an information network for providing new functionalities and digital business models.

Although IoT technology provides a fabulous opportunity for smart services across multiple sectors, it is not without security risks. Security concerns may be viewed as major roadblocks to the successful deployment of IoT-enabled smart services. This is because the IoT presents a unique set of access-control challenges at the perception layer due to the low power requirements of wireless sensor nodes, the memory limitations of the nodes to run complex encryption algorithms and the distributed nature of the extremely large number of sensor nodes and gateway devices required to create a system of systems for providing context-aware services. Not least, the complexity of the connectivity itself generates a risk, yet it is this connectedness that will drive the benefits of smart services in smart-city environments.

Wireless Sensor Nodes as Key Components of IoT Services

IoT-enabled new business models can gather data on a continuous basis from the contextual environment. Gathering data from various layers of business, technology and societal functions in unprecedented dimensions will be possible with the help of wireless sensor nodes on edge devices along with other technical means. According to NIST, sensors are one of the five core system primitives that "form the basic building blocks for a Network of 'Things' (NoT), including the Internet of Things (IoT)."*

However, wireless sensors, with immense potential to realize the benefits of the IoT, have some key limitations as of today. For example, these sensors have low computational power and low energy capacity. So they cannot store enough contextual data that are gathered from the defined environment, and running complex cryptographic algorithms for data security is also an issue. Wireless sensors cannot cover large areas for data gathering. Hence, for business and service needs, when a large number of sensors are deployed it creates a wireless mesh network having a free-scale topology.

The basic functionality of IoT-enabled services is depicted in Figure 7.1.

As shown here, the data collected by the sensor is sent wirelessly to an operations center that has a control system to monitor the relevance of collected data as per the contextual requirement. If the data is within the required range as desired for the business or service, then the control system allows the data to be stored for further application-based analytical processing and action. However, if the range for contextual data gathering requires a change, then the change instruction is

---

* Voas, J. (2016, July). Primitives and elements of internet of things (IoT) trustworthiness. Retrieved from http://csrc.nist.gov/publications/drafts/nistir-8063/nistir_8063_draft.pdf

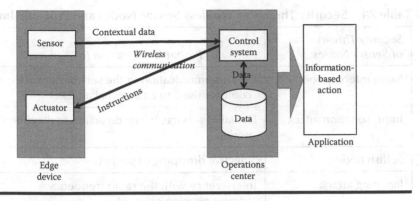

Figure 7.1   System design of an IoT application with wireless sensors.

passed on wirelessly to the actuator on the edge device from the control system and the sensors start collecting data as per the redefined range. The sensors can be remotely tracked, monitored and controlled.

Security Threats of Wireless Sensor Nodes

The scale and complexity of IoT-enabled services make it challenging to implement traditional security techniques for IoT services using wireless sensor nodes. The security of IoT-enabled smart services acquires a bigger dimension than traditional IT-enabled services due to the pervasive nature of these devices and services. The security threats to wireless sensor nodes and the probable impact of these threats* are depicted in Table 7.1.

Wireless sensor nodes can be aggregated in mesh networks to create a context-aware IoT service. The main characteristics of wireless mesh networks are: its decentralized structure, the dynamic nature of its network topology and its ease of access to a radio medium. As the attacker does not need to physically access the network, so a wireless sensor network is more vulnerable to external attacks. An attacker can listen to or transmit packets on a radio link from a distance in a wireless network. Also, an attacker can access the wireless mesh network from unpredictable locations at different times due to the mobility of these networks. The dependency of wireless sensor nodes on intermediate nodes for data transmission and the broadcasting method of the route to transmit data by these intermediate nodes make wireless sensor networks prone to various kinds of malicious attacks.

Adversaries can physically or remotely intercept and manipulate data captured by IoT sensor nodes. Data transmission from sensors and gateway devices can be

---

* Zhang, Y., Zheng, J., & Hu, H. (Eds.). (2008). *Security in wireless mesh networks.* CRC Press.

Table 7.1 Security Threats of Wireless Sensor Nodes and Probable Impact

Security Threats of Sensor Nodes	Probable Impact on IoT services
Node interception	Data communication in the sensor network can be compromised to an external attacker
Impersonation attack	Denial of service, traffic diversion to illegitimate nodes
Selfish nodes	Selective dropping of packets
Jamming attack	Interference with the radio frequency communication channel
Sinkhole attack	Attracts the communication data to form a black hole or selectively forwards data packets causing full or partial denial of service
Lack of authentication and authorization	Malicious node intrusion in the sensor network
Sybil attack	Communication channel hijack by malicious nodes, loss of data integrity
Network partitioning attack	Routing table disruption
Rushing attack	Route and data compromise by malicious nodes
Tunnel attack	Cooperative black hole

passively monitored if the encryption is not robust. Malicious nodes can be embedded in wireless sensor networks to access data and communicate with neighbor nodes.

The integrity of the routing protocol is a critical requirement for IoT-enabled services with wireless sensor nodes, because the wireless mesh networks depend on the cooperation of all legitimate sensor nodes to relay packets across the network. If there is any impersonation attack by malicious nodes that participate in the dynamic routing protocol and affect the choice of routers, then it can lead to multiple negative effects like the degradation and slowing of network performance, the denial of service to legitimate nodes in the network and the funneling of traffic through malicious nodes. These can ultimately lead to compromised IoT services that are not expected by users, and hence can breach user confidence and trust in the IoT service model.

The selfish behavior of sensor nodes in a wireless sensor network is another cause of concern for IoT services, as sensor nodes are dependent on each other for data forwarding. At a regular frequency, depending on their memory capacity, sensor nodes flush the contextual data to the gateway device for forwarding to the storage hosted in the cloud or in premise as per the architectural design of the service requirement. To ensure the right path of data transmission, a sensor node requests from other nodes in the network the route to the gateway device that can be reached in multiple hops through other legitimate sensor nodes in that mesh network. After responding to the route request, it might so happen that a selfish node either does not forward the data packets as per the route plan or drops packets. If all packets reaching the selfish node are dropped, then it can create a denial of service, and in the selective dropping of packets, the contextual data used in the application-based analytics function cannot provide effective service as desired from the IoT service's design. It is also hard to identify the root cause in such situations—whether it was a selfish node that caused a full or partial denial of service or if it was something else, like link failure or network congestion.

An intruder can join a network and carry out attacks on the wireless sensor nodes if there are no proper authentication and association mechanisms of the sensor nodes with the wireless sensor network. There can be hand-off delay of a wireless sensor node moving around the network as a mobile client, by switching its connectivity from one mesh router to another. During the hand-off, all ongoing transmissions of the client node are transferred to the new mesh router from the current mesh router. Hand-off delays can be a cause of concern if the business application requires real-time data transport from the sensor nodes without any drop of data packets during hand-off and delay.

Radio jamming by attackers external to the wireless sensor network can create interference that prevents data packets from flowing through the right route to the intended sensor node in the victim network. This is possible with wireless devices that can transmit strong signals to the victim network. The wireless device behaves as a malicious node and can attract all data packets from the sensor nodes in the network towards itself by claiming to have the right route for all route requests, even though it does not have the valid route to the destination. Due to the strength of the wireless signal of the malicious node, the legitimate nodes get distracted towards it and start sending data packets to the malicious node. The malicious node can redirect the data packets to the sink node, compromising confidentiality, or it can selectively drop the data packets leading to partial denial of service and thus create a sink-hole attack, or can drop all packets that it receives causing a black hole attack leading to a complete denial of service.

Malicious nodes can launch a rushing attack on the wireless sensor network by forwarding route request messages to a target node before the target node receives any route request from other legitimate nodes. When two or more malicious nodes collude together, these can create a tunnel to divert network traffic through that communication channel, preventing the data packets from reaching the sink node. This is a kind of cooperative black-hole attack. In a grey-hole attack, the malicious node drops data packets selectively to remain undetected for a longer duration of time. Also, there can be a Sybil attack when a malicious node creates multiple identities in the sensor network pretending to be a legitimate node each time to attract data traffic from other legitimate nodes in the network. Sometimes the malicious nodes can create a network-partitioning attack by disrupting routing tables to divide the network into non-connected partitions, resulting in denial of service.

In an IoT-enabled smart-service scenario like wireless point of sales at retail outlets, sensor-based healthcare devices using wireless communication, vending machines using wireless payment-card transactions or Bluetooth-enabled beacons, these threats can be a serious cause of concern to users, regulators and service providers if not addressed with caution. The consequences can have an enormous impact on security, privacy and trust, which are critical building blocks of these smart services and can be a hindrance to their reputation and the wider acceptance of these service models.

Security threats of IoT-enabled smart services at the sensor layer require adequate mitigation for reliable service offerings. For this, we need an effective approach for continuous monitoring of all functional wireless sensor nodes in order to identify and isolate any malicious or selfish behavior of a node before the smart service can be impacted. Smart-city service implementations will require a huge distribution of sensor nodes for a wider reach of the perception layer in smart-city service architecture. So it will be nearly impossible to physically track, round the clock, the health status and performance of the sensor nodes that are deployed for smart-service enablement. We have to adopt best practices to provide a continuous monitoring of sensor node performance and to ensure the security, privacy and reliability of IoT-enabled smart services. One of the approaches is to implement a risk-mitigation approach for IoT-enabled smart services, as discussed in the following section.

Threat-Mitigation Approach of Wireless Sensor Nodes for IoT Services

For IoT services with wireless sensor nodes, security is a prime necessity that needs to be ingrained in the IoT service model right from the service process design stage. We have to address security concerns with respect to the sanctity of contextual data gathered by the sensors at the perception layer, the reliability of communication from and to the wireless sensors in the wired and wireless networks of IoT services and the security of data stored in gateway and storage devices.

Smart-city services will require a huge number of IoT devices and wireless sensors interconnected in a free-scale topology. Smart city councils and IoT service providers should have accountability for IoT devices and services. There should be a service process owner as an overarching body to define the security policy, the network policy, data security and the IoT service asset management in order to provide end-to-end control of IoT devices, data and networks.

Several security concerns can come up regarding wireless sensors and wireless sensor networks when IoT-enabled services are implemented. To ensure that an IoT-enabled smart service is secured and trustworthy, the critical concerns provided in Table 7.2 need to be addressed in the smart service's design phase.

Smart city councils and IoT service providers can mitigate threats to wireless sensor nodes and ensure the security of IoT devices with the following steps:

1. **In an IoT system of systems, identify all wireless sensor assets.** This is an important first step to know what wireless sensor assets have been deployed for the smart service. For a smart-home automation system, there can be very few wireless sensors, but when a smart-city service is operational with sensors and networks across the city's landscape, there are many chances to lose track of all the wireless sensors being used for the smart service. When such untracked asset-sprawl creeps in, it becomes difficult for the service provider to ensure end-to-end security of the smart service.

2. **Identify the security threats of the wireless sensor assets at the configuration identifier level.** To prevent untoward scenarios as mentioned above, it is prudent for IoT service providers to set up a configuration management database (CMDB) and assign a configuration identifier (CI) for all wireless sensor assets with details of their purpose and approval for usage. Changes in the configurations of wireless sensor assets should be approved by the top management who responsible for providing a secured smart service.

3. **Security-threat assessment of wireless sensor assets.** A CI-level security-threat assessment of wireless sensor assets, on a continual basis, can help protect against security breaches. After identifying the threats, a security-threat registry should be prepared to provide visibility on the emerging threats.

4. **Develop appropriate security policies for threat mitigation.** Information technology is rapidly evolving, and security threats are also evolving. Hence, appropriate security policies should be invoked in practice to keep wireless mesh networks secured and abiding by the defined smart service for which

Table 7.2 Security Concerns for Wireless Sensors

Security Concerns
1. Can the data captured by sensors be compromised by other malicious nodes in the periphery?
2. Is the data from sensors copied to multiple locations?
3. How is the data from sensors being logged, and for what duration?
4. Is the data from sensors safe in transit?
5. Is the data sent from the sensors to the gateway device encrypted?
6. Who is sending control data to the actuator?
7. What data is sent back to the actuators?

they have been deployed. Security policies for wireless sensors and gateways should be constantly upgraded to address security challenges and threats.

5. **Maintain an IoT security life cycle for the smart service.** Smart-service providers should maintain an IoT security life cycle to plan and deploy the wireless sensors for their smart service, followed by the monitoring, detection, remediation and disposal of these wireless sensors by appropriate means. This is required to track inactive, unresponsive and erroneous wireless sensors and replace them to prevent any kind of complacency toward the security of the IoT devices and the smart service.

6. **Set up an IoT security incident and event management system to monitor and evaluate security incidents and new threats.** An IoT security incident and event management (SIEM) system, when enabled for a smart service, will provide the following features:
   - Continuous security monitoring of IoT infrastructure
   - Logging of events in the connected devices, networks and data
   - Analytics-based decision making
   - Detection of anomalies
   - Identification of new threats.

7. **Ensure periodic security assurance of IoT components and smart services.** A periodic security assurance of IoT components and smart services will help build a secured operational environment with resilience to external security threats. Wireless sensor nodes can fall prey to malicious activities instigated from inside by smart-service stakeholders and devices or from external attacks. The malicious insiders can be the smart-service operators and users, knowingly or unknowingly. A continuous monitoring and assurance function will help detect such threats early to prevent any kind of information-security compromise or critical smart-service lockdown. Auditing IoT components and smart services, including wireless sensors, will require the capability to understand emerging technologies, IoT business models, service designs, and IoT devices and sensors. A security audit questionnaire has been provided in Table 7.3 to address the security concerns of wireless sensor nodes for smart services.

IoT-enabled smart-service designs that utilize wireless sensor nodes require a multi-stakeholder approach for effective risk assessment and for the security of the smart services. Service providers and smart city councils have to play a leadership role with an overarching governance to realize the benefits of smart services with a perception of security and trust. Identifying the security threats to wireless sensor nodes on a continual basis with periodic information-security assurance activities as discussed in this chapter can provide a reliable and secured smart service.

Table 7.3 Security Audit Questionnaire for WSNs and Smart Services

1.	What are the authentication, authorization, association and accounting mechanisms established for the wireless sensor nodes (WSNs) used for the smart service?
2.	How many trust centers are there in the wireless network?
3.	Does the wireless mesh network use any kind of open-trust model?
4.	What is the access-control mechanism for the wireless sensor nodes?
5.	Are there any bridge networks in the smart-service architecture? Which bridging protocol is used? Is it secured from malicious attacks?
6.	Do the wireless sensor nodes authenticate a requesting WLAN client before servicing it?
7.	What routing protocol is used by the WSNs? Is it secured?
8.	Are the WSNs resilient to radiofrequency attacks and media-access attacks?
9.	What is the cryptographic technique used for WSN-to-WSN communication?
10.	Is there any message loss from the WSNs due to collision?
11.	Are there any malicious WSNs in the wireless mesh network?
12.	What is the physical health status of the WSNs?
13.	Are all WSNs tracked and registered in the CMDB?
14.	Is there any latency in the WSN-authentication process that can impact service continuity?
15.	Is any multi-layered security designed for the protocol layers?
16.	How does the downstream WSN authenticate the upstream WSN In the mesh network?
17.	How is trust established between the WSNs?
18.	Is it possible to auto-configure the WSNs?
19.	Is there any data owner for the data collected by the WSNs?
20.	Can the route discovery mechanism between the source and destination WSNs be intercepted and modified by external means?

(Continued)

Table 7.3 (Continued)   Security Audit Questionnaire for WSNs and Smart Services

21.	Is there any approved configuration-management process for the WSNs?
22.	Is there any authentication mechanism for the control data sent to the actuator?
23.	Where is the data from the WSNs stored? Who has access to that data?
24.	Are there any legal and regulatory requirements to be met for the IoT service? How are these requirements met?
25.	Are the smart service's users periodically informed or trained regarding how to use the smart service securely?

Suggested Reading

British Standards Institution. (2015). Smart city framework—Guide to establishing strategies for smart cities and communities. Retrieved from http://www.bsigroup.com/en-GB/smart-cities/Smart-Cities-Standards-and-Publication/PAS-181-smart-cities-framework/

Butun, I., Morgera, S. D., & Sankar, R. (2014). A survey of intrusion detection systems in wireless sensor networks. *IEEE communications surveys & tutorials*, *16*(1), 266–282.

Castillejo, P., Martínez, J. F., López, L., & Rubio, G. (2013). An internet of things approach for managing smart services provided by wearable devices. *International Journal of Distributed Sensor Networks*, *9*(2), 190813.

Chaudhuri, A. (2015). Address security and privacy concerns to fully tap into IoT's potential. Retrieved from http://www.tcs.com/SiteCollectionDocuments/White%20Papers/Address-Security-Privacy-Concerns-Fully-Tap-IoT-Potential-1015-1.pdf

Chaudhuri, A. (2016). Cyber risk mitigation for smart cities. Retrieved from http://www.tcs.com/SiteCollectionDocuments/White%20Papers/Cyber-Risk-Mitigation-Smart%20Cities-1015-1.pdf

Cloud Security Alliance. (2015). Identity and access management for the internet of things - summary guidance. Retrieved from https://downloads.cloudsecurityalliance.org/assets/research/internet-of-things/identity-and-access-management-for-the-iot.pdf

De Gante, A., Aslan, M., & Matrawy, A. (2014, June). Smart wireless sensor network management based on software-defined networking. In *2014 27th Biennial Symposium on Communications (QBSC)* (pp. 71–75). IEEE.

Eris, C., Saimler, M., Gungor, V. C., Fadel, E., & Akyildiz, I. F. (2014). Lifetime analysis of wireless sensor nodes in different smart grid environments. *Wireless networks*, *20*(7), 2053–2062.

Ferdoush, S., & Li, X. (2014). Wireless sensor network system design using Raspberry Pi and Arduino for environmental monitoring applications. *Procedia Computer Science*, *34*, 103–110.

Gubbi, J., Buyya, R., Marusic, S., & Palaniswami, M. (2013). Internet of things (IoT): A vision, architectural elements, and future directions. *Future generation computer systems, 29*(7), 1645–1660.

Guo, S., He, L., Gu, Y., Jiang, B., & He, T. (2014). Opportunistic flooding in low-duty-cycle wireless sensor networks with unreliable links. *IEEE Transactions on Computers, 63*(11), 2787–2802.

Hackmann, G., Guo, W., Yan, G., Sun, Z., Lu, C., & Dyke, S. (2014). Cyber-physical codesign of distributed structural health monitoring with wireless sensor networks. *IEEE Transactions on Parallel and Distributed Systems, 25*(1), 63–72.

Han, G., Jiang, J., Shu, L., Niu, J., & Chao, H. C. (2014). Management and applications of trust in wireless sensor networks: A survey. *Journal of Computer and System Sciences, 80*(3), 602–617.

He, D., Kumar, N., & Chilamkurti, N. (2015). A secure temporal-credential-based mutual authentication and key agreement scheme with pseudo identity for wireless sensor networks. *Information Sciences, 321*, 263–277.

International Electrotechnical Commission. (2014, June 31). Internet of things: Wireless sensor networks. Retrieved from http://www.iec.ch/whitepaper/pdf/iecWP-internet ofthings-LR-en.pdf

ITU-T FG-SSC. (2014). Focus group on smart sustainable cities. Retrieved from http://www.itu.int/en/ITU-T/focusgroups/ssc/Pages/default.aspx

Li, S., Zhao, S., Wang, X., Zhang, K., & Li, L. (2014). Adaptive and secure load-balancing routing protocol for service-oriented wireless sensor networks. *IEEE Systems Journal, 8*(3), 858–867.

Liu, Y., Dong, M., Ota, K., & Liu, A. (2016). ActiveTrust: secure and trustable routing in wireless sensor networks. *IEEE Transactions on Information Forensics and Security, 11*(9), 2013–2027.

Lu, H., Li, J., & Guizani, M. (2014). Secure and efficient data transmission for cluster-based wireless sensor networks. *IEEE transactions on parallel and distributed systems, 25*(3), 750–761.

Luo, X., Zhang, D., Yang, L. T., Liu, J., Chang, X., & Ning, H. (2016). A kernel machine-based secure data sensing and fusion scheme in wireless sensor networks for the cyber-physical systems. *Future Generation Computer Systems, 61*, 85–96.

MIT. (2014, July). Technology review business report: The internet of things. Retrieved from http://dspace.mit.edu/bitstream/handle/1721.1/86935/MIT-Technology-Review -Business-Report-The-Internet-of-Things.pdf

Miyazaki, T., Yamaguchi, S., Kobayashi, K., Kitamichi, J., Guo, S., Tsukahara, T., & Hayashi, T. (2014, February). A software defined wireless sensor network. In *2014 International Conference on Computing, Networking and Communications (ICNC)* (pp. 847–852). IEEE.

Mobile Working Group. (2015). Security guidance for early adopters of the internet of things (IoT). Retrieved from https://downloads.cloudsecurityalliance.org/whitepapers /Security_Guidance_for_Early_Adopters_of_the_Internet_of_Things.pdf

OWASP Internet of Things Project. (n.d.). Retrieved from https://www.owasp.org/index .php/OWASP_Internet_of_Things_Project

Patwari, N., Ash, J. N., Kyperountas, S., Hero, A. O., Moses, R. L., & Correal, N. S. (2005). Locating the nodes: Cooperative localization in wireless sensor networks. *IEEE Signal processing magazine, 22*(4), 54–69.

Perrig, A., Stankovic, J., & Wagner, D. (2004). Security in wireless sensor networks. *Communications of the ACM, 47*(6), 53–57.

Porambage, P., Schmitt, C., Kumar, P., Gurtov, A., & Ylianttila, M. (2014, April). Two-phase authentication protocol for wireless sensor networks in distributed IoT applications. In *Wireless Communications and Networking Conference (WCNC), 2014 IEEE* (pp. 2728–2733). IEEE.

Porter, M. E., & Heppelmann, J. E. (2014). How smart, connected products are transforming competition. *Harvard Business Review, 92*(11), 64–88.

Rahayu, T. M., Lee, S. G., & Lee, H. J. (2015). A secure routing protocol for wireless sensor networks considering secure data aggregation. *Sensors, 15*(7), 15127–15158.

Ransbotham, S. (2015). Ready or not, here IoT Comes. *MITSloan. Blog December, 22,* 2015.

Rezvani, M., Ignjatovic, A., Bertino, E., & Jha, S. (2015). Secure data aggregation technique for wireless sensor networks in the presence of collusion attacks. *IEEE Transactions on Dependable and Secure Computing, 12*(1), 98–110.

Roy, S., Conti, M., Setia, S., & Jajodia, S. (2014). Secure data aggregation in wireless sensor networks: Filtering out the attacker's impact. *IEEE Transactions on Information Forensics and Security, 9*(4), 681–694.

Tsilomitrou, O., Tzes, A., & Manesis, S. (2017, July). Mobile robot trajectory planning for large volume data-muling from wireless sensor nodes. In *2017 25th Mediterranean Conference on Control and Automation (MED)* (pp. 1005–1010). IEEE.

Tunca, C., Isik, S., Donmez, M. Y., & Ersoy, C. (2014). Distributed mobile sink routing for wireless sensor networks: A survey. *IEEE Communications Surveys & Tutorials, 16*(2), 877–897.

Turkanović, M., Brumen, B., & Hölbl, M. (2014). A novel user authentication and key agreement scheme for heterogeneous ad hoc wireless sensor networks, based on the internet of things notion. *Ad Hoc Networks, 20,* 96–112.

Xie, S., & Wang, Y. (2014). Construction of tree network with limited delivery latency in homogeneous wireless sensor networks. *Wireless personal communications, 78*(1), 231–246.

Younis, M., Senturk, I. F., Akkaya, K., Lee, S., & Senel, F. (2014). Topology management techniques for tolerating node failures in wireless sensor networks: A survey. *Computer Networks, 58,* 254–283.

Zhang, D., Li, G., Zheng, K., Ming, X., & Pan, Z. H. (2014). An energy-balanced routing method based on forward-aware factor for wireless sensor networks. *IEEE transactions on industrial informatics, 10*(1), 766–773.

Chapter 8

Managing Shared Risks in Interdependent Systems of Smart Cities

Intricate minglings of different uses in cities are not a form of chaos. On the contrary, they represent a complex and highly developed form of order.

Jane Jacobs

After reading this chapter you will be able to:

- Understand the interdependent systems of smart cities
- Understand the opportunities and risks of these interdependent systems
- Interpret the approach of mitigating these shared risks in smart cities
- Understand how standards and frameworks can be utilized for risk mitigation in smart cities.

Introduction

With an ever-increasing population in cities worldwide, it has become an enormous task to provide basic as well as enhanced facilities to city populations spanning all spheres of life. The number of urban residents is growing by approximately 60 million every year, and by 2050 it is estimated that people occupying just two percent of the world's land will consume about three-quarters of its resources.* To provide a better and connected experience, the concept of smart cities is being explored in various parts of the globe as a viable alternative to the current state of metropolitan life.

Information and communication technology (ICT) has enhanced business across the globe, first as an enabler and then as an integrated function of business. Now the smart city concept aims to enhance human experience in cities in the near future through ICT-enabled and IoT-embedded services. Also, smart cities are being projected as information-economy hubs for creating a knowledge society. And to make a city smart, there should be a risk-managed, secured underlying digital infrastructure that provides real-time data for all services and assisted living in the city.

There is a growing global trend of human settlements shifting from rural to urban, and hence cities are in an expansion mode. In this scenario, the concept of smart city is promising to provide enhanced city living experience with digitally enabled smart services for its citizens. New smart cities are also being designed to meet the needs of growing population, having smart, agile city services with interoperable Industrial Control Systems (ICS) and Information Technology Systems (ITS).

However, just like any other ICT initiative, smart cities also require proper policies and risk mitigation to ensure a secured, trustworthy environment for living, business, innovation, education and healthcare. Like any other ICT initiative, a smart city requires directions from top management, i.e. the city council, to keep itself efficient, safe and secured. A secured smart city requires risk-mitigation strategies to protect against any kind of cyber attack that can bring services to a halt. We are in the early stage of implementing smart cities in various countries, and there are no globally accepted standards and benchmarks on smart-city implementation and risk management that can be used as a reference. This makes the creation of secured smart cities even more challenging. Below, we discuss the shared risk-mitigation features that can be applied to the implementation and maintenance of smart cities.

Interdependent Systems in Smart Cities

Like existing cities, smart cities will have interdependent systems that build up their critical infrastructure. As shown in Figures 8.1 and 8.2, there can be multiple interdependent systems† in a city, which applies to a smart city as well. The

---

* Mitchell, S. et al. (2013). The internet of everything for cities: Connecting people, process, data, and things to improve the 'livability' of cities and communities. Retrieved from http://www.cisco.com/web/strategy/docs/gov/everything-for-cities.pdf
† Watson, J. (2014). The resilience of city systems. Retrieved from http://www.2014.csdmasia.net/IMG/pdf/CSDM_Jeremy_Watson.pdf

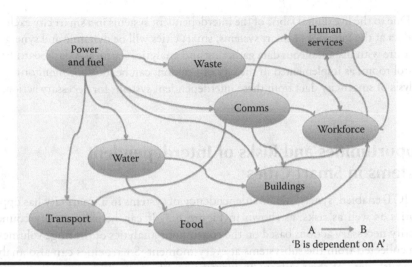

Figure 8.1 Interdependent systems in cities.

Watson, J. (2014). The resilience of city systems. Retrieved from http://www.2014 .csdmasia.net/IMG/pdf/CSDM_Jeremy_Watson.pdf

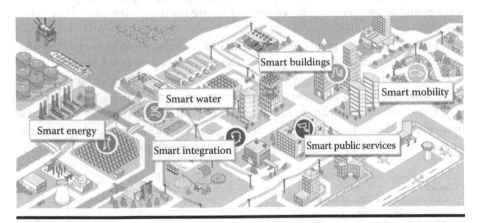

Figure 8.2 Systems of systems in cities.

Smart Cities. (2014). Retrieved from http://2014.gogreeninthecity.com/smart -cities.html

smart city has interdependent smart systems that can handle all major systems like water, energy generation and transmission, logistics, transport, garbage and waste disposal, green buildings, lighting systems, citizen-connected healthcare, online learning and education, people connected through smart phones, CCTVs, traffic systems and so on.

Due to the live digital fabric of the interdependent systems in a smart city exchanging data at the interface of smart systems, smart cities will be dynamic and synergistic in nature with instantaneous data gathering and analytics features. A dashboard-based control room, as implemented in the city of London, can be ideal for monitoring and analysis of smart-city data from these interdependent systems for necessary action.

Opportunities and Risks of Interdependent Systems in Smart Cities

The ICT-enabled, synergistic interdependency of systems in a smart city has opportunities as well as risks, as shown in Figure 8.3. It can help smart city councils identify necessary actions based on the continuous analytics of the huge volumes of data collected from the subsystems at every moment. Smart cities can also analyze the health data of their citizens to identify health scares, like virus attacks at an early stage, in order to take necessary action. Data integration in smart cities can also be utilized for mapping the energy efficiency of buildings, mapping social data for crime prevention, monitoring flood situations, public consultations and trend analysis and city development in areas like housing, education, transport, medical services and employment. Other opportunities are the decreasing cost of operation and improving flexibility.

However, this synergistic interdependency of the subsystems in a smart city also carries risks in operation. If one system fails to provide relevant information to the next connected system, it might create chaos in a smart city. For example, a failure in the smart traffic-management database server can disrupt the seamless operation of the smart transport-management system, as vehicles might continue to clog a particular highway junction instead of being diverted to a ring road, due to the unavailability of information from the traffic-management system.

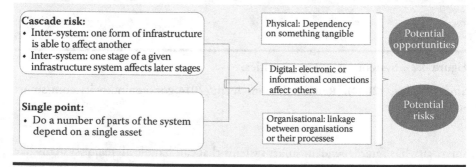

Figure 8.3 Opportunities and risks of interdependent systems in smart cities.

Watson, J. (2014). The resilience of city systems. Retrieved from http://www.2014 .csdmasia.net/IMG/pdf/CSDM_Jeremy_Watson.pdf.

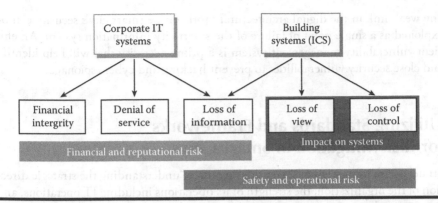

Figure 8.4 Risks arising from compromised interdependent systems in smart cities. Boyes, H. (2013). Resilience and cyber security of technology in the built environment. Retrieved from http://www.cpni.gov.uk/documents/publications/2013 /2013063-resilience_cyber_security_technology_built_environment.pdf?eps language=en-gb

Also, for smart healthcare services in a smart city, there are tremendous opportunities to digitally monitor the health of diabetes and heart disease patients through data exchange on a 24/7 basis from pacemakers and smart glucose monitors implanted in the patients' bodies directly to citizens' health-monitoring systems in smart hospitals. In such a critical service, if the Internet connectivity of the glucose monitor or pacemaker is breached or the authenticity of the data cannot be verified, then the patients' lives might be at risk.

As shown in Figure 8.4, a smart-city system of systems will have a mesh of interconnected industrial control systems (ICSs) and IT systems (ITS). Any vulnerability in the security architecture of the ICSs or IT systems in a smart city can open a floodgate for disaster in terms of virus attacks, security breaches, compromises of sensitive data and lockdowns of critical services. While the ICSs will have safety and operational risks, the IT systems will have financial and reputational risks. Gathering and analyzing real-time data with supervisory control and data acquisition is of prime importance for smart-system security in smart cities to prevent security failures and a complete lockdown of services.

Necessity of Risk Mitigation in Smart Cities

Due to the large number of devices that are expected to be connected in a smart city's digital infrastructure, enhanced security management for gateway devices is of prime importance to prevent data breaches or leakage when one system feeds data to another system in the interdependent cyber systems. There can be no compromise of the security of the digital infrastructure of a smart city, because any weak link in the digital architectural fabric of the smart city's security can be exploited as a single point of failure of the smart-city information system. An efficient vulnerability-management system is a prime necessity that will help identify and close security vulnerabilities to prevent hacking and cyber espionage.

Utilizing Standards and Frameworks for Risk Mitigation in Smart Cities

Strategic risk mitigation of smart cities requires "understanding the strategic direction of the organization, the strength of its operations including IT operations, and the threats that may prevent it from achieving the goals and objectives."[*] A smart city council should be operated like a modern-day enterprise with specified goals and objectives, including cyber defense of the smart-city cyber systems. The smart city council should continuously evaluate, direct and monitor the security of the smart-city cyber systems and the privacy of the citizens.

There is no specific standard yet that can be consulted by smart city councils to implement or maintain a secured, interconnected cyber system for their smart city. But various existing standards and frameworks for governance and control can be utilized by smart city councils. For example, the BSI PAS 181 framework[†] can be utilized by smart city councils to develop a smart-city framework and to establish security and privacy strategies for smart-city dwellers. Cyber-risk assessments should be performed on a regular basis to understand new risks, and all threats leading to these risks should be identified for proper countermeasures. The OCTAVE[‡] risk-assessment framework can be utilized to identify the information assets and the corresponding containers to assess risks qualitatively on an ongoing basis and to secure them.

The smart-city concept is a multi-stakeholder ownership approach to a dynamic ecosystem created with integrated, live, interdependent cyber systems. Effective risk assessment and management of the interconnected cyber systems is a prime necessity for ensuring the security of smart-city services. Smart city councils have to play a leadership role with an overarching governance authority to manage the shared risks of the interdependent smart systems to help citizens realize the benefits of smart-city living with a sense of security, trust and confidence.

---

[*] Chaudhuri, A. (2011). Enabling effective IT governance: Leveraging ISO/IEC 38500: 2008 and COBIT to achieve business–IT alignment. *Edpacs*, *44*(2), 1–18.

[†] BSI. (2014). PAS 181 smart city framework. Retrieved from http://www.bsigroup.com/en-GB/smart-cities/Smart-Cities-Standards-and-Publication/PAS-181-smart-cities-framework/

[‡] Caralli, R.A. et al. (2007). Introducing OCTAVE Allegro: Improving the Information Security Risk Assessment Process. Retrieved from http://resources.sei.cmu.edu/asset_files/TechnicalReport/2007_005_001_14885.pdf

Suggested Reading

Abbott-Donnelly, I., & Lawson, H. B. (2017). *Creating, analysing and sustaining smarter cities: A systems perspective.* College Publications.

Anthopoulos, L. (2015, August). Defining smart city architecture for sustainability. In *Proceedings of the 14th IFIP Electronic Government (EGOV) and 7th Electronic Participation (ePart) Conference 2015.* Presented at the 14th IFIP Electronic Government and 7th Electronic Participation Conference 2015, IOS Press (pp. 140–47).

Aurigi, A. (2016). No need to fix: Strategic inclusivity in developing and managing the smart city. In *Digital futures and the city of today.* Intellect.

Cohen, B. (2013). Smart city wheel. Retrieved from https://www.smart-circle.org/smartcity/blog/boyd-cohen-the-smart-city-wheel/

CPNI. (2015). Good practice guide – process control and scada security. Retrieved from http://www.cpni.gov.uk/Documents/Publications/2008/2008031-GPG_SCADA_Security_Good_Practice.pdf

Gordon, L. W., & McAleese, G. W. (2017). Resilience and risk management in smart cities. Retrieved from https://cip.gmu.edu/2017/07/06/resilience-risk-management-smart-cities/

Guthrie, P. K. (2012). Infrastructure and resilience. Retrieved from https://www.gov.uk/government/uploads/system/uploads/attachment_data/file/286993/12-1310-infrastructure-and-resilience.pdf

ISO/IEC. (2008). Corporate Governance of Information Technology. Switzerland: ISO. Retrieved from http://www.iso.org/iso/catalogue_detail?csnumber=51639

Korolov, M. (2015). Attacks against industrial control systems double. Retrieved from http://www.itworld.com/article/2911634/attacks-against-industrial-control-systems-double.html

Kitchin, R. (2015). Making sense of smart cities: Addressing present shortcomings. *Cambridge Journal of Regions, Economy and Society, 8*(1), 131–136.

Li, Y., Dai, W., Ming, Z., & Qiu, M. (2016). Privacy protection for preventing data over-collection in smart city. *IEEE Transactions on Computers, 65*(5), 1339–1350.

Little, R. G. (2002). Controlling cascading failure: Understanding the vulnerabilities of interconnected infrastructures. *Journal of Urban Technology, 9*(1), 109–123.

Nam, T., & Pardo, T. A. (2011, September). Smart city as urban innovation: Focusing on management, policy, and context. In *Proceedings of the 5th international conference on theory and practice of electronic governance* (pp. 185–194). ACM.

NCIIPC. (2013). Guidelines for protection of national critical information infrastructure. Retrieved from http://perry4law.org/cecsrdi/wp-content/uploads/2013/12/Guidelines-For-Protection-Of-National-Critical-Information-Infrastructure.pdf

OECD. (2015). The OECD recommendation on digital security risk management for economic and social prosperity. Retrieved from http://www.oecd.org/sti/ieconomy/digital-security-risk-management.pdf

Office, C. (2012). A Summary of the 2012 Sector Resilience Plans. Retrieved from https://www.gov.uk/government/uploads/system/uploads/attachment_data/file/271349/Summary-2012-Sector-Resilience-Plans.pdf

Rinaldi, B. P. (2001). Identifying, Understanding, and Analyzing Critical Infrastructure Interdependencies. *IEEE Control Systems Magazine, 21*(6), 11–25.

Sanchez, L., Muñoz, L., Galache, J. A., Sotres, P., Santana, J. R., Gutierrez, V., Ramdhany, R., Gluhak, A., Krco, S., Theodoridis, E., & Pfisterer, D. (2014). SmartSantander: IoT experimentation over a smart city testbed." *Computer Networks, 61,* 217–238.

Sterbenz, J. P. (2016, June). Drones in the smart city and IoT: Protocols, resilience, benefits, and risks. In *Proceedings of the 2nd Workshop on Micro Aerial Vehicle Networks, Systems, and Applications for Civilian Use* (pp. 3–3). ACM.

Sterbenz, J. P. (2017, September). Smart city and IoT resilience, survivability, and disruption tolerance: Challenges, modelling, and a survey of research opportunities. In *2017 9th International Workshop on Resilient Networks Design and Modeling (RNDM)* (pp. 1–6). IEEE.

Townsend, A. M. (2013). Smart Cities: Big Data, Civic Hackers, and the quest for a new utopia. (First ed.). London: W.W. Norton & Company Ltd.

Tragos, E. Z., Angelakis, V., Fragkiadakis, A., Gundlegard, D., Nechifor, C. S., Oikonomou, G., & Gavras, A. (2014, March). Enabling reliable and secure IoT-based smart city applications. In *2014 IEEE International Conference on Pervasive Computing and Communications Workshops (PERCOM Workshops)* (pp. 111–116). IEEE.

Vanolo, A. (2014). Smartmentality: The smart city as disciplinary strategy. *Urban Studies*, *51*(5), 883–898.

Wang, P., Ali, A., & Kelly, W. (2015, August). Data security and threat modeling for smart city infrastructure. In *2015 International Conference on Cyber Security of Smart Cities, Industrial Control System and Communications (SSIC)* (pp. 1–6). IEEE.

Chapter 9

Smart City Governance

The governance framework is there to encourage the efficient use of resources and equally to require accountability for the stewardship of those resources. The aim is to align as nearly as possible the interests of individuals, corporations and society.

Sir Adrian Cadbury

After reading this chapter you will be able to:

- Understand the necessity of governance in smart cities
- Gain an insight on cyber threats in smart cities
- Understand the role of smart city councils in smart city governance
- Understand the role of citizens in smart city governance
- Interpret the relevance of governance standards in smart cities
- Gain an insight on governance and policies to address cyber defense in smart cities.

Introduction

A smart city will deal analytically with huge volumes of data that will be generated from the communication between subsystems (machine to machine, or M2M) and from the interaction with the citizens (machines to people, or M2P, and people to machines, or P2M), as shown in Figure 9.1.

Private and sensitive information directly obtained, intercepted or inferred from data flows in a smart-city system of systems can be a cause of deep concern with regard to the safety and security of smart-city services and citizens. Health informatics will

Figure 9.1   Interactive M2M, P2M and M2P communications in a smart city.

also be an integrated service in a smart city, and there should be absolute security and privacy of citizens' health data. In such a scenario, the basic need of the citizens will be the assurance of secured and effective transmission of data, including private health data, in the smart-city network. Any incident of data breach or data loss will jeopardize the citizens' sense of security and trust in smart-city services and living. Some other information-security concerns in a smart city can be the interception of wireless data in transit between sender and receiver, the leakage of confidential information through traps laid by motivated externals and the implantation of viruses and Trojans in devices like sensors. Cloud-based information services and data storage in smart cities can be compromised through hacking and related subversive activities.

The security of IoT installations should be ensured by the smart city council to prevent any physical attack or infiltration. Identity-management mechanisms should be employed for user and device authentication at every interface of the smart systems in a smart city. IP-security mechanisms should also be ensured to provide a highly robust and resilient smart-city system of systems.

The Role of Smart City Councils in the Risk Mitigation of Smart Services

The risk mitigation of smart services in smart cities requires a detailed understanding of several factors. These include the design and architecture of the smart services, the IT-infrastructure service teams' capabilities, and cyber threats. A smart city council should operate like a modern-day enterprise with specific goals and objectives that include planning for cyber defense of smart services and building capabilities for emergency response. Ensuring the security of networks and sensors is of prime importance to provide trustworthy smart services in a city.

The smart city council should secure IoT installations to prevent any physical attack or infiltration. Identity-management and device-authentication mechanisms should be employed at every interface of a smart system. The effective security of networks and sensors should also be ensured to provide a highly robust and resilient smart-city ecosystem. Digital forensic capabilities should be integrated with the smart-city cyber architecture right from the design phase. This will help city councils gather any evidence of untoward incidents and take appropriate action. Gathering and analyzing real-time data with supervisory control and data acquisition (SCADA) is necessary to prevent security failures and a complete lockdown of critical services.

Building Resilient Systems

As a smart city grows, the interconnections of systems and interdependencies of smart services increase manifold. This makes the smart systems more vulnerable to cyber attacks. Smart city councils should therefore aim to design a risk-resilient digital architecture for the interdependent systems of their smart cities. This should possess the adaptive capability to arrest anomalies in the nascent stage, and lock down a subsystem without disturbing other live components, ensuring uninterrupted service delivery.

Building resilient, interdependent systems in smart cities will ensure that city councils are well prepared to handle cyber emergencies and can restore any impacted services quickly. An effective cyber-resilience strategy has to be designed, tested and implemented to protect cyber assets in case of any eventuality. Smart-city business-continuity planning (BCP) is an effective risk management initiative that can help smart city councils ensure the security and availability of smart services. Periodic BCP drills should be conducted, audited, and documented for ready reference during criticalities. This will enable smart cities to take a recovery-oriented approach toward risk management.

Adopting International Standards

The security standards and risk mitigation strategies currently being used to secure IT systems (ITSs) may not be adequate to safeguard the interdependent systems in smart cities. However, ISO22301:2012, the 'International Standard for Societal Security—Business Continuity Management Systems,'* can be adopted by smart city councils to prevent the disruption of smart services. Proper communication management is critical for smart cities to respond to critical cyber threats and other exigencies. Communication channels with pre-identified specific points of contact should be defined, documented, and regularly updated, and documents should be made available to all stakeholders of a smart city, including residents.

---

* ISO. (2008). Societal security -- business continuity management systems -- requirements. Retrieved from https://www.iso.org/standard/50038.html

Performing System-Impact and Interdependency Analysis

Periodic system-impact analysis should be performed to identify risks posed to critical interdependent systems and interconnected services, with appropriately defined recovery-time and recovery-point objectives. Smart cities should also have secured data receivers and data storage to continuously collect and store data generated from the industrial control system (ICS) and ITS components for analysis, analytics decision making and incident response.* The stored data should be periodically backed up as a precautionary measure to deal with any cyber-emergency situation. As a precautionary measure, data flow from control systems can be channelized using data diodes to prevent data contamination.

Smart city councils should devise a component protection strategy to identify critical components of the interdependent smart systems for agile risk analysis. Preliminary system-interdependency analysis should be conducted to understand the requirements for information continuity at the system interfaces and to identify the critical components that enable the flow of vital information. This should be followed by a probabilistic interdependency analysis to manage risks of high-fidelity interdependent systems like smart grids and smart health-monitoring systems for senior citizens and critical patients. This type of analysis can be very helpful in enhancing the resilience of critical systems in a smart city. The CPNI Good Practice Guide for Process Control and SCADA Security† can be utilized by smart city councils to ensure the security and trustworthiness of the interdependent systems in their smart cities. It provides a framework based on industry good practice from process control and IT security and focuses on seven key themes:

1. Understand the business risks
2. Implement secure architecture
3. Establish response capabilities
4. Improve awareness and skills
5. Manage third-party risks
6. Engage projects
7. Establish ongoing governance.

---

* Chaudhuri A. Managing Cyber Risks in Smart City System of Systems. Chapter 5; Abbott-Donnelly, I., & Lawson, H. B. (2017). *Creating, Analysing and Sustaining Smarter Cities: A Systems Perspective*. College Publications.
† CPNI. (2015). Good practice guide – process control and scada security. Retrieved from http://www.cpni.gov.uk/Documents/Publications/2008/2008031-GPG_SCADA_Security_Good_Practice.pdf

FIVE QUESTIONS TO THE GLOBAL LEADER

Chuck Benson
Assistant Director for IT
University of Washington

1. What is your vision of a secured-IoT ecosystem for smart living?

I don't envision a completely secured IoT ecosystem. I don't believe that is attainable. However, I do believe that cities and institutions can develop and mature a risk-mitigated IoT ecosystem that will provide cities and institutions with the highest levels flexibility, adaptability, performance and resilience that are attainable.

Such a risk-mitigated IoT ecosystem would have at least these three components:

- The city or institution has clear expectations of what the IoT system will do once deployed and can manage to those expectations
- IoT systems vendors are highly vetted with preference given to vendors that financially share in the risk of system success
- Language and conceptual frameworks around IoT systems continuously evolve.

The first of these ecosystem components is that cities and institutions know what they want. They need this self-knowledge in order to measure performance, to know what risks that they are incurring with IoT systems implementations, and to set expectations for IoT systems providers. An important subset of that set of expectations is determining what cyber security and risk management deliverables are required.

Mature and well-managed vendor relationships comprise the second component. Because of the nature of networks, devices and systems, cities and institutions are increasingly intertwined. Clear demarcation of boundaries is gone. Performance expectations must be well communicated and agreed to and the vendor choices of preference are those vendors and providers that financially share in the risk of successful IoT system implementation.

Finally, the component underlying all of this is language development across providers, city/institutional consumers and departments/organizations within an institution or city. Currently, there is limited language and limited precedent for these kinds of interactions, problem solving and information sharing. Language and conceptual frameworks around IoT systems need to be continuously developed, evolved and mutually shared so that procurement, operations and governance can continuously mature.

2. **What should be the steps to develop a trustworthy IoT supply chain for smart cities?**

Two components of a trustworthy IoT supply chain are provenance-review standardization and an IoT-specific vendor vetting process.

Regarding provenance-review standardization, I was really influenced by a presentation video that I stumbled on a few years ago by Andrew 'bunnie' Huang and Sean 'xobs' Cross at the 2013 Chaos Computer Congress. The talk was about how they hacked a microcontroller on an SD memory card. (I think most of us would not even suspect that memory cards have embedded CPU's on them.) In the course of the presentation, Huang describes vast bins of memory cards of ranging quality, size and performance in the market of Huaqiangbei in Shenzhen, China. He talks about card relabeling as a common practice to adjust for sub-performing cards as well as card factories that have very few access controls regarding what files are written to cards and chips and how they are configured.

Prior to his description, I had the naive notion that all memory cards, integrated circuits, and other electronic components were developed in tightly controlled and regulated manufacturing and distribution environments with extensive record keeping on every device.

Many buy from this kind of loosely controlled electronics market because the components are so cheap compared to a highly regulated manufacturer. And IoT devices have a lot of these kinds components.

IoT devices can have hardware and software components from multiple providers, such as:

■ Sensing hardware—the physical parts that interact with the environment
■ Device drivers for sensing hardware
■ Memory access software
■ Control software
■ Networking software—TCP/IP stack, ether, wireless, etc.
■ Local data-analysis software

- Web software—customer and provider facing as well as web services
- Encryption software
- Device-management software
- Upstream system controllers and data aggregators
- Others.

For all of these, where do these components come from? Who makes them? Do they have separately sourced subcomponents themselves? If so, what is the source of those subcomponents?

This is a lot to track and manage, and much of it may well not be trackable or manageable. By developing a standardized approach to reporting provenance across all of the multiple components, we can begin to mitigate (but not eliminate) the risk of some of these components not being what their label says.

The second component is a vendor vetting process for IoT systems. This process is unique for IoT, because while there may be vendor/provider vetting processes for traditional enterprise systems, there are not the same processes for IoT systems that involve a part or all of the enterprise.

Developing a vetting process for IoT systems specifically is critical for expectations of operational effectiveness and institutional/city cyber-risk mitigation.

3. **How can we make universities smarter with IoT technology?**

Because universities are often like small- to medium-sized cities, they can benefit from appropriately selected, implemented and managed IoT systems just as cities can. Examples include safety systems, traffic-management systems, energy-management systems, environmental monitoring such as air quality, structural monitoring such as buildings and roads, lighting control, waste management, and many others.

Universities can actually be more complex than cities because of their broad range of activities and services. In addition to city functions such as police departments, transportation management, facilities management, housing, utility management and others, universities also have several other services related to their teaching, research, athletics and service missions. Examples include:

- **Scientific research systems**. IoT-based environmental monitoring and control systems have rapidly rising usage on campus. Regardless of discipline, almost all science seeks to establish an environmental baseline of some sort to do research. IoT systems are a great fit for that objective.
- **Athletics**. Transforming the fan-based experience is a new and rapidly growing industry. Universities, with their football and baseball

stadiums, basketball arenas and other venues are major opportunities to evolve the IoT-driven fan-based experience.

▪ **Teaching and learning.** There are rich opportunities for IoT systems in learning-management systems, classroom interaction, immersive learning, laboratory experiences and others.

4. **What are the key success factors for successful IoT systems implementations in smart cities?**

Two broad components to analyze to determine a successful IoT system implementation are return on investment (ROI) and changes to the city's cyber-risk profile.

The first, determining the ROI is often more difficult than it would appear to be. The simple definition is—does the system do what was expected for the actual dollars spent on initial investment, and which continue to be spent on operating costs? A tricky part can be answering the initial question, "What was the expectation of this IoT system?" Several factors go into this expectation question:

▪ The demographics of the city's citizenry are broad. The potential variety of incomes, education levels, city neighborhoods and regions and other factors can serve to create a corresponding spectrum of expectations. Minimally, the assumption should not be made that all citizens will have the same expectations of the system and perceive potential benefits in the same manner.

▪ Because IoT systems span many departments within a city—from planning and budgeting, capital development, central IT, facilities management, departmental IT groups, many vendors and contractors and others—another spectrum of expectations regarding IoT system implementation, operation and support can also exist.

▪ As Big Data and the derivative products and services of Big Data become increasingly pervasive across a city's population, attributes and nuances of interpreting and mediating data become more visible in practice. Researchers Brittany Fiore-Gartland and Gina Neff have proposed the term "data valences," and suggest six data valences, including actionability of data, self-evidence of data, and data transparency, in their paper "Communication, Mediation and the Expectations of Data: Data Valences Across Health and Wellness Communities."*

---

* Fiore-Gartland, B., & Neff, G. (2015). Communication, mediation, and the expectations of data: data valences across health and wellness communities. *International Journal of Communication, 9*, 19.

Determining the actual investment and costs of an IoT systems deployment can also be challenging. Part of the reason for this is that purchase decisions can be made in one part of the city's government, e.g. a planning/budgeting department, but the costs of implementation and long-term support of the system are actually borne by other departments, e.g. central IT, the transportation department, and perhaps others. Support costs, to include operations and repair, can be difficult to capture across multiple departments with their own potentially unique methods of operation and accounting. Particularly disruptive is when supporting departments are not aware of these IoT systems purchases and don't have time to plan and allocate resources for the support.

Regarding the second component of determining IoT system-implementation success in a city, the simple version of the question is, "Did the city's cyber-risk profile get worse in the course of deploying and operating the new IoT system?"

IoT systems offer plenty of opportunity to make cyber risk worse for a city. The raw volume of deployed, networked computing devices (the 'T' in IoT) creates many more potential points of exploitation for an adversary, whether direct or indirect. These devices may be deployed by the thousands, tens of thousands or more in the city.

Worsening the problem is the potential for poor management of these devices. This, in turn, has at least two contributing factors. One factor is that because of the nature of their sensing and broad geographic deployments, these networked computing devices tend to be out of sight, out of mind, and are often not considered as networked computers—as traditional desktops, servers and even virtual machines are. Because of this, they can easily miss being included in the city's risk picture.

Another factor in the risk of poorly implemented and managed devices comes from vendor relationships and the institution's maturity in setting appropriate expectations of IoT systems and installation vendors. The city must mandate detailed configuration deliverables for IoT systems deployments. Examples include:

■ Ensure the default login/password are changed on every installed device
■ Ensure all unnecessary services are disabled on every installed device
■ Deliver a test plan that ensures compliance with the above
■ Deliver documentation of every device location, IP and MAC address, device configuration, software/firmware version numbers, etc.

These example deliverables as well as others need to be a part of the contract and/or purchasing language.

Successful implementation minimally requires evaluating ROI and post-implementation cyber-risk analysis, and both require new processes, approaches and resourcing for cities.

5. **What is your opinion regarding developing information-risk resilience in smart cities?**

It is a critical objective and I also believe it is a very immature concept in most cities and institutions. The advent and rapid growth of the cloud, IoT, Big Data and the interdependencies between them has created a complexity aspect for city management that is not fully comprehended yet and possibly never will be. We will never nail down and fully understand all of the existing and rapidly evolving complexity in a city, but it is important that we know and adjust our behavior to this rapid change in complexity.

This delta between the cyber criminal's or nation state operator's awareness of this rapidly evolving complexity and lack of awareness of the same in cities and institutions creates the margin, operating space and opportunity that the adversary needs.

Ensuring Citizen Compliance

Citizens of smart cities will play a crucial role in ensuring the security of interdependent systems—from cyber- as well as physical-security perspectives. Citizens with smart devices who are critical points in the cyber-system framework can be targeted by attackers and hackers to gain entry into the system. This can be done through social engineering, spam emails, data streaming and other malicious activities. To prevent this, smart city councils should develop policies and procedures for establishment, maintenance and operation of secure smart services. And cyber-awareness programs should be made mandatory for citizens, with penalties for non-compliance.

Governance and Policies to Address Cyber Defense in Smart Cities

Effective governance and policies are required for the security of smart services in smart cities and to ensure the privacy of their citizens. An architectural framework has to be put in place by smart city councils taking into consideration the essential components that are required to "share data, process transactions and secure critical systems."* Although there is a lack of industry standards for smart and secured cities, smart city councils can take a lead in this matter by instituting standards and practices that will be suitable for particular cities. There cannot be a single solution template for the security and privacy of all smart cities. They have to be considered on a case-by-case basis, because every smart city will be unique in its architecture, patterns, culture and the behavior of its citizens.

Smart city councils should be the overarching body taking up a governing role for ensuring the security, safety and privacy of their smart city. They should be responsible for implementing suitable policies for the implementation, maintenance and usage of secured smart services.

The notion of the security and privacy of smart cities should be incorporated by smart city councils right from the initial design phase of the city's information architecture. Hasty designs and shortcuts to meet stringent deadlines for smart-service implementation create the possibility for errors and can create gateways for hacks that can prove detrimental to the smart-living concept.

Contingency measures have to be kept ready by smart city councils to counter any security incidents in the smart-city system of systems that can jeopardize the seamless functioning of smart-city services. There can be huge repercussions from a single incident of security breach at any of the communication points of a smart city inter-network of systems. Smart city councils should continuously identify and assess the biggest security risks of their smart city's assets and should take appropriate action for risk mitigation. The 12 questions in Table 9.1 below have to be brought up repeatedly in smart city council board meetings, and should have definite answers with board-level acceptance.

Utilizing Standards and Frameworks for Cyber Defense in Smart Cities

The strategic governance of a smart city requires "understanding the strategic direction of the organization, the strength of its operations including IT operations, and the threats that may prevent it from achieving the goals and objectives."† A smart city council should be operated like a modern day enterprise with specified goals and objectives, including cyber defense of the smart-city cyber systems.

There is no single, specific standard that can be consulted by smart city councils to implement or maintain a secured, interconnected cyber system for the smart city. But various existing standards and frameworks for governance and control can be

---

* Townsend, A.M. (2013). *Smart cities: Big data, civic hackers, and the quest for a new utopia.* (First ed.). London: W.W. Norton & Company Ltd.
† Chaudhuri, A. (2011). Enabling effective IT governance: Leveraging ISO/IEC 38500: 2008 and COBIT to achieve business–IT alignment. *EDPACS 44*(2), 2–18.

Table 9.1   12 Questions for a Smart City Council's Board for Smart City Governance

1.	When and how can smart systems fail? What will the failover policy be for the smart subsystems?
2.	Do we have enough knowledge of the consequences of a security breach and how it can affect the critical, interconnected subsystems?
3.	Are we doing the right things right in implementing the security and privacy of the smart systems network of the city? How can we build a just, social and sustainable smart city for our citizens?
4.	How can we ensure the authenticity of smart devices and users on the smart-city network?
5.	What level of access privileges should we provide to the various levels of smart devices and users in the smart-city cyber system?
6.	Should we opt for a private cloud for accessing critical data from the smart systems of the city?
7.	Are enough service-continuity drills and security drills being performed to confirm that the smart, interconnected systems are appropriate for use by the citizens?
8.	What will be the role of the citizens in the maintenance, enhancement and continuous improvement of smart-city services and experience? Should we follow a top-down structural approach, where citizens will only be end users? Or should they be allowed to participate and co-innovate in enhancing the resiliency of the smart-city cyber systems?
9.	How can we utilize the huge volume of data that is generated at every moment in the smart city network for a better smart-living experience?
10.	What will be the policy for assuring the security and privacy of smart-city services? What will be the regulations for compliance to data privacy and cyber security? What are the key assurance metrics we need to ensure a secured, smart cyber system? What will be the frequency of these assurance activities?
11.	Who will be the legal owner of the smart services and the data generated from these smart subsystems?
12.	What will be the policy to establish the supply chain for implementation, execution and maintenance of the smart-city cyber framework and services? How can we ensure the security and privacy of the smart-city cyber systems with third-party service providers in the supply chain?

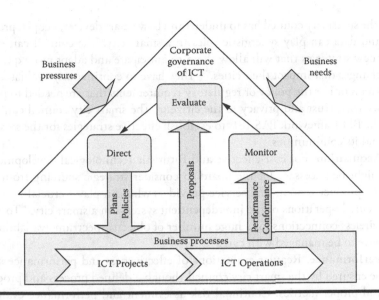

Figure 9.2   ISO/IEC 38500:2008 Model for Corporate Governance of IT.

ISO. (2008). Corporate governance of information technology. Retrieved with permission, from http://www.iso.org/iso/catalogue_detail?csnumber=51639.

utilized by smart city councils, like the ISO/IEC 38500:2008 standard, as shown in Figure 9.2.

(The figure from ISO/IEC 38500:2008, Corporate governance of information technology, is reproduced with the permission of the International Organization for Standardization (ISO). This standard can be obtained from any ISO member and from the website of the ISO Central Secretariat at the following address: www .iso.org. Copyright remains with ISO.)

A smart city council can continuously evaluate, direct and monitor the security of the smart-city cyber systems and the privacy of the citizens based on the following six governance principles from the ISO/IEC 38500:2008 standard:

■ **Responsibility.** A smart city council should develop a security strategy for the smart-city cyber systems and provide authority to identified teams and individuals for management and execution of the necessary duties. There should be a clear distinction between the responsibilities of the governance board and the management team. A proper governance structure should be set up in the organization with adequate reporting and communication channels for two-way communication—top down and bottom up, and the performance by each role should be monitored periodically using well-defined indicators.

■ **Strategy.** The key strategic requirements for smart cities are data governance and ethical exploitation of data for the benefit of the cities' residents.

The smart city council has to understand how smart devices, people, processes and data can play synergistic roles in a smart city. The council can explore "new strategies that will allow them to anticipate and adapt more quickly to changes that impact their cities."* They have to continuously evaluate, direct and monitor the policy or regulatory requirements that are needed to provide security, trust and privacy to the citizens. The smart city council can utilize the BSI framework PAS 181† to establish effective strategies for the smart city and its communities.

▪ **Acquisition**. For cost-effective and futuristic technological developments, it might be necessary for the board to consider strategic sourcing from external product vendors and service providers who can play a crucial role in the secured operations of the interdependent systems in a smart city. "To ensure wireless connections for a huge number of objects, spectrum availability will have to be managed with care."‡

▪ **Performance**. Regular evaluations of effectiveness and performance should be ensured by the smart city council though a defined process and procedure and proper metrics. Continual risk assessment and performance evaluation of the cyber systems is mandatory to ensure security and trust in a smart-city cyber system.

▪ **Conformance**. The smart city council should ensure secured performance of the smart systems and their ethical use. Policies should be established for the confidentiality, integrity, availability, security and privacy of authentic users and data in the smart-city cyber system. Legal, regulatory conformance of the smart-city cyber systems should be ensured by the smart city council through internal audits, security and vulnerability tests and external audits.

▪ **Human behavior**. The smart-city cyber system will be designed, developed and used by humans. Appropriate use of the cyber systems should be ensured by the smart city council through proper policies, procedures, streetwise initiatives for the citizen on the street and appropriate learning and skill development initiatives for all stakeholders who occupy various roles in the smart-city cyber fabric.

Understanding and evaluating risks in smart-city systems require a pragmatic approach to cyber-risk management due to the high level of interconnectedness of smart services and the rapidly evolving nature of the systems. With smart-city services projected to grow rapidly over the next few years, there is a clear need for

---

* Mitchell, S. et al. (2013). The internet of everything for cities. Retrieved fromhttp://www .cisco.com/web/strategy/docs/gov/everything-for-cities.pdf
† BSI. (2014). PAS 181 smart city framework. Retrieved from http://www.bsigroup.com/en-GB /smart-cities/Smart-Cities-Standards-and-Publication/PAS-181-smart-cities-framework/
‡ European Commission. (2015). The internet of things. Retrieved from http://ec.europa.eu /digital-agenda/en/internet-things

smart city councils to focus on mitigating security concerns. Incorporating risk mitigation and developing strong security strategies in the initial planning stages will enable smart city councils to provide safe, secure, and reliable services to their citizens.

Suggested Reading

Alawadhi, S., Aldama-Nalda, A., Chourabi, H., Gil-Garcia, J. R., Leung, S., Mellouli, S., Nam, T., Pardo, T. A., Scholl, H. J. & Walker, S. (2012, September). Building understanding of smart city initiatives. In *International Conference on Electronic Government* (pp. 40–53). Springer, Berlin, Heidelberg.

Anthopoulos, L. G., Janssen, M., & Weerakkody, V. (2015, May). Comparing Smart Cities with different modeling approaches. In *Proceedings of the 24th International Conference on World Wide Web* (pp. 525–528). ACM.

Bakıcı, T., Almirall, E., & Wareham, J. (2013). A smart city initiative: The case of Barcelona. *Journal of the Knowledge Economy, 4*(2), 135–148.

Carli, R., Deidda, P., Dotoli, M., & Pellegrino, R. (2014, September). An urban control center for the energy governance of a smart city. In *Emerging Technology and Factory Automation (ETFA), 2014 IEEE* (pp. 1–7). IEEE.

Casbarra, C., Amitrano, C. C., Alfano, A., & Bifulco, F. (2014, November). Smart city governance for sustainability. In *HASSACC–Human and Social Sciences at the Common Conference, 2*(1). EDIS–Publishing Institution of the University of Zilina.

Castelnovo, W., Misuraca, G., & Savoldelli, A. (2016). Smart cities governance: The need for a holistic approach to assessing urban participatory policy making. *Social Science Computer Review, 34*(6), 724–739.

Chourabi, Hafedh, Taewoo Nam, Shawn Walker, J. Ramon Gil-Garcia, Sehl Mellouli, Karine Nahon, Theresa A. Pardo, and Hans Jochen Scholl. (2012). Understanding smart cities: An integrative framework. In *2012 45th Hawaii International Conference on System Science, HICSS 2012*, Maui, HI (pp. 2289–2297). IEEE.

Clohessy, T., Acton, T., & Morgan, L. (2014, December). Smart City as a Service (SCaaS): A future roadmap for e-government smart city cloud computing initiatives. In *Proceedings of the 2014 IEEE/ACM 7th International Conference on Utility and Cloud Computing* (pp. 836–841). IEEE Computer Society.

Coe, A., Paquet, G., & Roy, J. (2001). E-governance and smart communities: A social learning challenge. *Social Science Computer Review, 19*(1), 80–93.

Dameri, R. P., & Rosenthal-Sabroux, C. (2014). Smart city and value creation. In *Smart city* (pp. 1–12). Springer International Publishing.

di Bella, E., Odone, F., Corsi, M., Sillitti, A., & Breu, R. (2014). Smart Security: Integrated systems for security policies in urban environments. In *Smart city* (pp. 193–219). Springer International Publishing.

Gabrys, J. (2014). Programming environments: Environmentality and citizen sensing in the smart city. *Environment and Planning D: Society and Space, 32*(1), 30–48.

Gargiulo, C., Pinto, V., & Zucaro, F. (2013). EU Smart city governance. *Tema–Journal of Land Use, Mobility and Environment, 6*(3), 356–370.

Goldsmith, S., & Crawford, S. (2014). *The responsive city: Engaging communities through data-smart governance*. San Francisco: John Wiley & Sons.

Kitchin, R. (2014). The real-time city? Big data and smart urbanism. *GeoJournal, 79*(1), 1–14.

Kitchin, R. (2015). Making sense of smart cities: Addressing present shortcomings. *Cambridge Journal of Regions, Economy and Society, 8*(1), 131–136.

Kitchin, R. (2016). Getting smarter about smart cities: Improving data privacy and data security. Data Protection Unit, Department of the Taoiseach, Dublin, Ireland.

Lee, J. H., & Hancock, M. (2012). *Toward a framework for smart cities: A comparison of Seoul, San Francisco and Amsterdam.* Innovations For Smart Green Cities: What's Working, What's Not, What's Next, Oberndorf Event Center. Retrieved from http://docplayer.net/8185388-Toward-a-framework-for-smart-cities-a-comparison-of-seoul-san-francisco-amsterdam.html

Lee, J. H., Phaal, R., & Lee, S. H. (2013). An integrated service-device-technology roadmap for smart city development. *Technological Forecasting and Social Change, 80*(2), 286–306.

Lee, J., & Lee, H. (2014). Developing and validating a citizen-centric typology for smart city services. *Government Information Quarterly, 31*, S93–S105.

Little, R. G. (2002). Controlling cascading failure: Understanding the vulnerabilities of interconnected infrastructures. *Journal of Urban Technology, 9*(1), 109–123.

Lombardi, P. (2011). New challenges in the evaluation of smart cities. *Network Industries Quarterly, 13*(3), 8–10.

Lombardi, P., Giordano, S., Farouh, H., & Yousef, W. (2012). Modelling the smart city performance. *Innovation: The European Journal of Social Science Research, 25*(2), 137–149.

Meijer, A. (2016). Smart city governance: A local emergent perspective. In *Smarter as the new urban agenda* (pp. 73–85). Springer International Publishing.

Meijer, A., & Bolivar, M. (2013, September). Governing the smart city: Scaling-up the search for socio-techno synergy. Paper presented at 2013EGPA Conference Proceedings, Edinburgh, Scotland.

Meijer, A., & Bolívar, M. P. R. (2016). Governing the smart city: A review of the literature on smart urban governance. *International Review of Administrative Sciences, 82*(2), 392–408.

Nam, T., & Pardo, T. A. (2011, September). Smart city as urban innovation: Focusing on management, policy, and context. In *Proceedings of the 5th International Conference on Theory and Practice of Electronic Governance* (pp. 185–194). ACM.

Neirotti, P., De Marco, A., Cagliano, A. C., Mangano, G., & Scorrano, F. (2014). Current trends in smart city initiatives: Some stylised facts. *Cities, 38*, 25–36.

OECD. (2015). *Digital Security Risk Management for Economic and Social Prosperity: OECD Recommendation and Companion Document.* Retrieved from http://www.oecd.org/sti/ieconomy/digital-security-risk-management.pdf

Ojo, A., Curry, E., & Janowski, T. (2014). Designing next generation smart city initiatives-harnessing findings and lessons from a study of ten smart city programs. ECIS.

Pan, J. G., Lin, Y. F., Chuang, S. Y., & Kao, Y. C. (2011, May). From governance to service-smart city evaluations in Taiwan. In *2011 International Joint Conference on Service Sciences* (pp. 334–337). IEEE.

Paskaleva, K. A. (2009). Enabling the smart city: The progress of city e-governance in Europe. *International Journal of Innovation and Regional Development, 1*(4), 405–422.

Paskaleva, K. A. (2011). The smart city: A nexus for open innovation? *Intelligent Buildings International, 3*(3), 153–171.

Popescu, G. H. (2015). The economic value of smart city technology. *Economics, Management and Financial Markets, 10*(4), 76.

Rinaldi, B. P. (2001). Identifying, understanding, and analyzing critical infrastructure interdependencies. *IEEE Control Systems Magazine, 21*(6), 11–25.

Scholl, H. J., & AlAwadhi, S. (2016). Smart governance as key to multi-jurisdictional smart city initiatives: The case of the eCityGov Alliance. *Social Science Information, 55*(2), 255–277.

van Zoonen, L. (2016). Privacy concerns in smart cities. *Government Information Quarterly, 33*(3), 472–480.

Vanolo, A. (2014). Smartmentality: The smart city as disciplinary strategy. *Urban Studies, 51*(5), 883–898.

Viale Pereira, G., Cunha, M. A., Lampoltshammer, T. J., Parycek, P., & Testa, M. G. (2017). Increasing collaboration and participation in smart city governance: A cross-case analysis of smart city initiatives. *Information Technology for Development, 23*(3), 526–553.

Walravens, N., & Ballon, P. (2013). Platform business models for smart cities: From control and value to governance and public value. *IEEE Communications Magazine, 51*(6), 72–79.

Chapter 10

Internet of Things and Regulations

Regulations force people to do better.

Jay Leno

After reading this chapter you will be able to:

- Understand the impact of regulations on IoT applications
- Understand the regulatory requirements of the EU GDPR for IoT businesses
- Gain an insight on the UN IGF net-neutrality principle from the IoT's perspective
- Understand the regulatory impact of blockchain technology on the IoT
- Interpret the strategic alignment of user-privacy needs for IoT applications.

Introduction

The IoT being an emerging technology, the regulatory and compliance landscape is also evolving with the growth in use of the 'Things' and related concerns on various aspects that can impact users and society. The regulatory landscape for the IoT will also be quite broad because of the potential applications of this technology in various business domains. With the growing popularity of IoT devices and smart services and increased adoption across the globe, we are also seeing a growth in user concerns on information security and privacy resulting from the association with these smart devices and services. While governments are developing new regulations or utilizing existing ones to regulate the IoT, we are also seeing new regulations like the EU GDPR that aim to address the user's concerns from the use or ownership of IoT devices and smart services. Regulations like the EU GDPR can have a significant impact on IoT businesses if they are not aligned. The IoT relies heavily on the Internet as a key communication network, and we are also seeing norms being designed in various countries for maintaining or modifying the need for network neutrality that can impact IoT adoption and usage and smart services. Blockchain is another emerging technology that can have a significant impact on IoT-enabled smart applications in finance and other emerging smart services. The regulatory impacts on the IoT are discussed in this chapter considering these various aspects.

The EU General Data Protection Regulation and the Internet of Things

The General Data Protection Regulation (GDPR),* approved by the EU Parliament and enforceable from May 25, 2018, has been designed to protect and empower the data privacy of all EU citizens. The GDPR text is extensive, with 99 Articles (Table 10.1) and 173 Recitals (Table 10.2). User data from IoT applications and services will also come under the purview of the GDPR. As we have seen in Chapter 10, privacy concerns can arise if IoT devices and smart services gather user data that are not relevant to the offered service.

Incidents of data breaches and privacy concerns are being reported in the media about IoT devices and services like smart retail, smart toys, smart TVs, smart watches, smart homes and so on. For example, the German Federal Network Agency† has banned smart watches for kids, as these have been found to have remotely accessible microphones that can allow spying on children wearing the watches and the people around them. Using an app, the children's parents can also listen to the environment surrounding the child without being noticed.

Smart-home scenarios can pose a privacy conundrum, as described by the European Union Agency for Network and Information Security:

> Privacy issues in smart homes are not limited to confidentiality and access control. Smart home sensors in particular will generate a large amount of highly personal data about activities within the home. The multiple streams of data combined together in a smart home system create the possibility of deeper contextual background and reveal patterns of behaviour of the inhabitants. The visibility of the smart home

---

* Regulation, G. D. P. (2016). Regulation (EU) 2016/679 of the European Parliament and of the Council of 27 April 2016 on the protection of natural persons with regard to the processing of personal data and on the free movement of such data, and repealing Directive 95/46. Official Journal of the European Union (OJ), 59, 1–88. Retrieved from https://publications.europa.eu /en/publication-detail/-/publication/3e485e15-11bd-11e6-ba9a-01aa75ed71a1/language-en
† Lomas, N. (2017, November). Germany bans kids' smartwatches that can be used for eavesdropping. Retrieved from https://techcrunch.com/2017/11/20/germany-bans-kids-smartwatches -that-can-be-used-for-eavesdropping/

occupant is increased by the large network of third parties who may be involved in providing smart home functionality. Smart home functions may have serious impacts upon privacy of the person, privacy of behaviour and action, privacy of communication, privacy of data and image, privacy of location, and privacy of association. . . . Smart home systems may include embedded features that are opaque to the user, and do not inform the user about the status of their operation.*

As per Article 83 of the EU GDPR, organizations not complying with the GDPR will have to pay a fine of either up to 4% of annual global turnover or 20 million euros, whichever is higher. There can also be a tiered approach to fines for a variety of reasons, like a fine of 2% for not having records in order, for not notifying the supervising authority and the data subject during a data breach or for not conducting a privacy-impact assessment. Both the data controller and the data processor will come under the ambit of these rules.

According to Article 7 of the GDPR, a data subject's consent is necessary for the processing of personal data by an IoT application, and service functions and the data controller must be able to demonstrate it on request. The IoT service provider has to ensure that the data subject has provided consent to the processing of personal data for obtaining the smart-service output.

If personal data is collected from the data subject for any smart service, then as per Article 13 of the GDPR, information has to be provided to the data subject regarding what metadata is gathered for processing that can impact user privacy.

Article 15 of the GDPR emphasizes the right of access of the data subject to personal data that is being used for any IoT function. The data subject can request information regarding the purpose of processing the personal data, any categorization of that data, to whom the personal data have been disclosed and the planned duration of storage of that data.

The context of Article 17 is the 'right to erasure' that can be exercised by the data subject when withdrawing from a smart service or device. If personal data has been processed by unlawful means, then the data subject can exercise the right to restriction of processing as per Article 18(b) or can object to the processing of personal data (Article 21). Automated decision making and user profiling are key concerns from smart services that collect contextual metadata and utilize user-behavior analytics to profile the likes and dislikes of users. Utilizing Article 22, the end user can object to automated decision making and profiling without consent.

As per Article 25 and Recital 78, appropriate technical and organizational measures are necessary to protect IoT data by design and by default. Measures like

---

* Barnard-Wills, D., Marinos, L., & Portesi, S. (2014, December). Threat Landscape and Good Practice Guide for Smart Home and Converged Media. European Union Agency for Network and Information Security (ENISA). Retrieved from https://www.enisa.europa.eu/publications /threat-landscape-for-smart-home-and-media-convergence/at_download/fullReport

Table 10.1 Specific GDPR Articles Relevant to IoT Data Protection

Regulatory Context	GDPR Article
Conditions for consent	7
Information to be provided where personal data are collected from the data subject	13
Right of access by the data subject	15
Right to erasure (the 'right to be forgotten')	17
Right to restriction of processing	18
Right to object	21
Automated individual decision making, including profiling	22
Data protection by design and by default	25
Processor	28
Records of processing activities	30
Security of processing	32
Notification of a personal data breach to the supervisory authority	33
Communication of a personal data breach to the data subject	34
Data-protection impact assessment	35
Designation of the data protection officer	37
Codes of conduct	40
Certification	42
Transfers on the basis of an adequacy decision	45
Transfers subject to appropriate safeguards	46

Source: Regulation, G. D. P. (2016). Regulation (EU) 2016/679 of the European Parliament and of the Council of 27 April 2016 on the protection of natural persons with regard to the processing of personal data and on the free movement of such data, and repealing Directive 95/46. Official Journal of the European Union (OJ), 59, 1–88. Retrieved from https://publications .europa.eu/en/publication-detail/-/publication/3e485e15-11bd-11e6-ba9a -01aa75ed71a1/language-en

Table 10.2 Specific GDPR Recitals Relevant for IoT Data Protection

Regulatory Context	Recital No.
Application of pseudonymization to personal data	28
Personal data concerning health	35
Processing of personal data	39
Availability, authenticity, integrity and confidentiality of stored or transmitted personal data	49
Principles of fair and transparent processing	60
The right to have personal data rectified and the 'right to be forgotten'	65
Personal data processed for the purposes of direct marketing	70
Automated processing of personal data evaluating the personal aspects	71
Appropriate technical and organizational measures	78
Clear allocation of responsibilities	79
Ensuring appropriate levels of security, including confidentiality	83
Notifying personal-data breach to supervisory authority without undue delay	85
Communication of personal-data breach from controller to data subject	86
Data-protection impact assessment	90
Personal-data transfer from the EU to third countries or international organizations	101

Source: Regulation, G. D. P. (2016). Regulation (EU) 2016/679 of the European Parliament and of the Council of 27 April 2016 on the protection of natural persons with regard to the processing of personal data and on the free movement of such data, and repealing Directive 95/46. Official Journal of the European Union (OJ), 59, 1–88. Retrieved from https://publications .europa.eu/en/publication-detail/-/publication/3e485e15-11bd-11e6-ba9a -01aa75ed71a1/language-en

IoT 'Privacy by Design' are necessary for data protection. Article 32 and Recitals 49 and 83 emphasize the appropriate level of security for the data being processed. Proper encryption is necessary for personal data in storage or in motion. 'Pseudonymization' of personal data captured by IoT devices and smart services is necessary to protect against making inferences and user profiling. As per Article 35, the data controller should perform a data-protection impact assessment (DPIA) on the operational data in the smart service. A designated data-protection officer can be consulted, as per Article 37, by the data controller for the DPIA to ensure that relevant safeguards are in place.

All processing of personal data after transfer from the EU region to third countries or international organizations, including onward transfers of personal data to another third country or international organization (Article 44, Recital 101), should ensure appropriate safeguards (Article 46) and comply with adequacy decisions (Article 45).

IoT Privacy-Alignment Strategy for Regulatory Compliance

IoT device manufacturers, smart-service providers and smart city councils should adopt a stakeholder approach for IoT data protection and privacy considering users at the core of the product or service design. Alignment of user-privacy needs is a primary necessity to provide trustworthy IoT services. This requires a three-step IoT privacy-alignment strategy, as described below:

Step 1: Perform Privacy Inquisition and Analysis

In this step, the data-privacy concerns of users should be identified and mapped with the IoT business needs of the organization or the smart-service needs of the smart city council. This inquisition is necessary to understand the gaps in service or product design with respect to the end-user expectations. A privacy inquisition model is shown in Figure 10.1, which can be utilized to map the privacy concerns of IoT users with the needs of IoT businesses and to identify the non-aligning factors for the next step.

Step 2: Conduct IoT Privacy-Impact Assessment

Conduct a privacy-impact assessment (PIA) on the IoT information flows and identify the privacy risks covering all aspects of the data gathered and processed, the consent received and the regulatory requirements. A well planned PIA can help to identify the business impact and associated costs from the existing product or service design, the financial impact of penalties for non-compliance to regulations and the financial impact on brand and trustworthiness for any reputational damage in cases of data breach or loss.

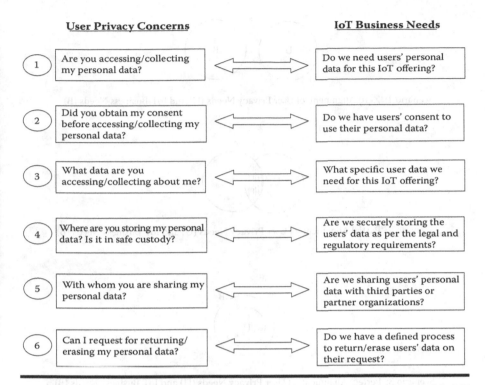

Figure 10.1   Privacy inquisition model for aligning user privacy needs with IoT business needs.

Chaudhuri, A. (2016). Internet of things data protection and privacy in the era of the general data protection regulation. *Journal of Data Protection & Privacy, 1*(1), 64–75.

Step 3: Transition Toward IoT Privacy Maturity

Based on the PIA outcome, necessary processes and technical improvements should be planned to transition to an advanced level of privacy maturity that complies with the needs of businesses, users and regulators. Appropriate privacy-enhancing technologies should be identified for the operating environment that can enhance the privacy-preserving capabilities and assist in mitigating privacy risks of the organization.

The above steps should be performed periodically under the supervision of a privacy governance board endowed with overarching authority and represented by a data controller, a data protection officer and key members of the organization's board of directors.

The principles of IoT 'Privacy by Design' and a novel IoT privacy framework have been provided in Appendix A with application use-case in a smart-home scenario.

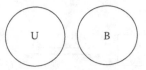

Scenario 1: 'Zero Alignment' of User Privacy Needs (U) and IoT Business Needs (B)

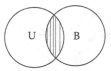

Scenario 2: 'Partial Alignment' of User Privacy Needs (U) and IoT Business Needs (B)

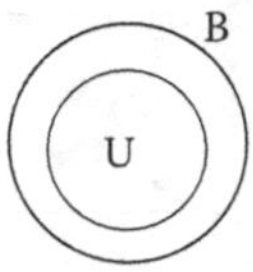

Scenario 3: 'Perfect Alignment' of User Privacy Needs (U) and IoT Business Needs (B)

Figure 10.2   State transitions to perfect privacy alignment.

Chaudhuri, A. (2016). Internet of things data protection and privacy in the era of the general data protection regulation. *Journal of Data Protection & Privacy, 1*(1), 64–75.

The Internet of Things and Net Neutrality

Although IoT devices can be configured to function in a license-free wireless frequency in most global locations, the Internet will remain a key medium of communication for 'Things' across the globe. Depending on the criticality of the smart service offered, a higher level of quality of service (QoS) standards might become a necessity for the projected billions of IoT devices that will be operational in the coming years. Regulatory frameworks and policies at national or regional levels on network usage by 'Things' can have a significant impact on the IoT ecosystem and smart-city services. The interplay of 'Internet Access Services,' QoS standards, 'Transparency in Network Traffic' and 'Specialized IoT Offerings'* will be the influencing factors for providing normality in the communication of 'Things.'

---

* The Committee on Communications Policy. (2010, November). NETWORK TRAFFIC MANAGEMENT AND THE EVOLVING INTERNET. WHITE PAPER. Retrieved from https://ieeeusa.org/wp-content/uploads/2017/07/IEEEUSAWP-NTM2010.pdf

Table 10.3   Preamble to the UN IGF Policy Statement on Network Neutrality

1. The Internet should be open, secure and accessible to all people.
2. Network Neutrality plays an instrumental role in preserving Internet openness; fostering the enjoyment of Internet users' human rights; promoting competition and equality of opportunity; safeguarding the generative peer-to-peer nature of the Internet; and spreading the benefits of the Internet to all people.
3. Managing Internet traffic in a transparent and non-discriminatory manner compatible with the Network Neutrality Principle serves the interests of the public by preserving a level playing field with minimal barriers to entry and by providing equal opportunity for the invention and development of new applications, services and business models.
4. Competition among broadband networks, technologies and all players of the Internet ecosystem is essential to ensure the openness of the Internet.

Source: Internet Governance Forum. (2015, November). Outcome Document on Network Neutrality. Retrieved from https://www.intgovforum.org/cms /documents/igf-meeting/igf-2016/833-dcnn-2015-output-document/file

The term 'network neutrality' was coined by Columbia University media-law professor Tim Wu in 2003. According to the United Nations Internet Governance Forum (UN IGF) Policy Statement on Network Neutrality, "Network Neutrality is the principle according to which Internet traffic is treated without discrimination, restriction or interference regardless of its sender, recipient, type or content, so that Internet users' freedom is not restricted by unreasonably favoring or disfavoring the transmission of specific Internet traffic."*

The preamble of this policy statement considers four important factors, as shown below in Table 10.3, to ensure alignment with the network neutrality principle.

Network-neutral treatment of IoT services can create a 'win-win' situation in realizing the benefits of the IoT and smart services. The internet-service infrastructure has to expand in tandem with the growth of 'Things' in order to prevent issues like clogged networks and breakdowns of smart services.

Blockchain and IoT

Blockchain technology can play a crucial role in IoT M2M transactions through digital contracts and distributed ledgers that are secured, accessible and auditable.

---

* Internet Governance Forum. (2015, November). Outcome Document on Network Neutrality. Retrieved from https://www.intgovforum.org/cms/documents/igf-meeting/igf-2016/833-dcnn -2015-output-document/file

FIVE QUESTIONS FOR THE GLOBAL LEADER:

Dr. Luca Belli
Senior researcher at the Center for Technology and Society, FGV Brazil
Founder and co-chair of the Dynamic Coalition on Network Neutrality
United Nations Internet Governance Forum

1. **Should IoT applications come under the purview of net neutrality?**

    Yes indeed. By mandating non-discriminatory treatment of Internet traffic, net neutrality is instrumental to the development of an open IoT environment in which operators cannot discriminate traffic coming from or directed to specific devices. Of course, this does not mean that reasonable traffic-management measures can be utilized to differentiate traffic based on the protocol utilized by the entire class of IoT applications, should this be necessary to cope with congestion and ensure that the technical requirement of the class of applications are available. The goal of net neutrality is not to hinder innovation, but rather to avoid specific innovators being unduly discriminated against—for instance by throttling or blocking traffic coming from or directed to specific devices built by specific producers or used by specific users. Should the specific IoT service require some technical features that are not possible on the Internet (e.g. enhanced security or guaranteed quality), the specific IoT service could also be offered as a specialized service. The 'Model Framework on Network Neutrality'* provides some basic clarifications on specialized services.

    * UN Internet Governance Forum Dynamic Coalition on Network Neutrality. Model Framework on Network Neutrality. Retrieved from http://www.networkneutrality.info/sources.html

2. **Is the UN IGF Policy Statement on Network Neutrality applicable for IoT applications and smart-city services?**

    Yes. The 'Policy Statement on Network Neutrality'* can be applied to IoT applications and smart-city services. If network infrastructure has sufficient capacity, non-discriminatory traffic management does not impede the development of the IoT or smart cities. If network infrastructure has limited capacity, then operators (and/or municipalities, in the case of smart-city services) should invest in the expansion of network capacity rather than implementing discriminatory traffic management. The IoT and smart-city services may require dedicated infrastructure, when their technical requirement cannot be achieved on the Internet. If this latter case is objectively proven, then the IoT and smart-city services should be provided as specialized services with dedicated infrastructure.

* UN Internet Governance Forum Dynamic Coalition on Network Neutrality. Policy Statement on Network Neutrality. Retrieved from https://www.pdf-archive.com/2018/01/21/dcnn-outcome-document -net-neutrality-policy-statement/preview/page/1/

3. **There is a mention of 'reasonable traffic management' in the UN IGF Policy Statement on Network Neutrality. Can IoT network traffic be reasonably managed without violating the net-neutrality principle for priority services like critical healthcare IoT applications?**

The Policy Statement foresees the possibility to deviate from net neutrality as long as it is necessary and proportionate to specific circumstances, such as prioritizing emergency services in the case of unforeseeable circumstances or force majeure. If critical healthcare or IoT applications require features that are not possible on the Internet-access service, then they should be provided as specialized services.

4. **Considering cloud-based IoT applications that can be offered as well as accessed from any geographic region, should we have specifically designed international regulations for complying with net-neutrality norms?**

Yes, this would be an ideal scenario. It would allow all internet users to enjoy the same choice of services and/or to become IoT service providers and provide their service globally with no risk to be blocked or throttled. Such an international framework would be both pro-innovation, pro-competition and pro-consumer.

5. **Predictions from various studies indicate an exponential growth in the usage of IoT-enabled services globally by 2025. How can we ensure conformance to the net-neutrality principle in the future world of 'Things'?**

By having clear rules; by empowering regulators with the resources and tools (including technological tools) necessary to monitor their implementation; and, critically, by involving users in such monitoring activities, e.g. offering them applications to monitor traffic shaping in a decentralized fashion through their connected devices (e.g. connected cars, wearables, etc.). Multi-stakeholder cooperation is increasingly essential to both developing and implementing Internet policy. In this

perspective, the Policy Statement clearly foresees that "All individuals and stakeholders should have the possibility to contribute to the detection, reporting and correction of violations of the Network Neutrality Principle."
* Ibid.

By utilizing blockchain, digital transactions between IoT systems can be made available to the public for transparency in operation.* In a system of 'Things,' blockchain technology can ensure the autonomy of handshaking and data exchange between IoT devices and smart systems without the need for any certifying authority or human interference.

Regulations will play an important role in such scenarios to ensure the acceptable blockchain operating criteria in IoT applications. The security and privacy of data in IoT blockchain will also be important considerations of regulatory frameworks. We expect to see a flurry of activities in the regulatory front to make IoT blockchain legitimate, secured, trusted and mainstream.

Regulations will have a significant impact on IoT technology and smart applications. Regulators and policy makers have to ensure that the innovative efforts of IoT are not nipped in the bud due to cautionary measures and that stakeholder concerns are addressed appropriately to keep the IoT ecosystem innovative, socially accountable and disruptive.

Suggested Reading

Almeida, V. A., Doneda, D., & Monteiro, M. (2015). Governance challenges for the internet of things. *IEEE Internet Computing, 19*(4), 56–59.

Arias, O., Wurm, J., Hoang, K., & Jin, Y. (2015). Privacy and security in internet of things and wearable devices. *IEEE Transactions on Multi-Scale Computing Systems, 1*(2), 99–109.

Brass, I., Tanczer, L., Carr, M., & Blackstock, J. (2017). Regulating IoT: Enabling or disabling the capacity of the internet of things? *Risk & Regulation, (16)*33, 12–15.

Caron, X., Bosua, R., Maynard, S. B., & Ahmad, A. (2016). The Internet of Things (IoT) and its impact on individual privacy: An Australian perspective. *Computer Law & Security Review, 32*(1), 4–15.

Coleman, S., Pothong, K., Vallejos, E. P., & Koene, A. (2017). The internet on trial: How children and young people deliberated about their digital rights. Report available at http://casma. wp.horizon.ac.uk/wp-content/uploads/2016/08/Internet-On-Our -Own-Terms.pdf.

---

* Compton, J. (2017, June). How Blockchain Could Revolutionize The Internet Of Things. Retrieved from https://www.forbes.com/sites/delltechnologies/2017/06/27/how-blockchain -could-revolutionize-the-internet-of-things/#5b3ec7146eab

D'Orazio, C. J., Choo, K. K. R., & Yang, L. T. (2017). Data exfiltration from internet of things devices: iOS devices as case studies. *IEEE Internet of Things Journal, 4*(2), 524–535.

Dai, W., Qiu, M., Qiu, L., Chen, L., & Wu, A. (2017). Who moved my data? Privacy protection in smartphones. *IEEE Communications Magazine, 55*(1), 20–25.

Dorri, A., Kanhere, S. S., Jurdak, R., & Gauravaram, P. (2017, March). Blockchain for IoT security and privacy: The case study of a smart home. In *IEEE International Conference on Pervasive Computing and Communications Workshops (PerCom Workshops), 2017* (pp. 618–623). IEEE.

Edwards, L. (2016). Privacy, security and data protection in smart cities: A critical EU law perspective. *European Data Protection Law Review, (2)*1, 28–58.

European Commission. (n.d.). IoT privacy, data protection, information security. Retrieved from http://ec.europa.eu/information_society/newsroom/cf/document.cfm?action =display&doc_id=1753

Gong, T., Huang, H., Li, P., Zhang, K., & Jiang, H. (2015, December). A medical healthcare system for privacy protection based on IoT. In *Seventh International Symposium on Parallel Architectures, Algorithms and Programming (PAAP), 2015* (pp. 217–222). IEEE.

Hacker, P. (2017). Personal data, exploitative contracts, and algorithmic fairness: Autonomous vehicles meet the internet of things. *International Data Privacy Law, 7*(4), 266–286.

Henze, M., Hermerschmidt, L., Kerpen, D., Häußling, R., Rumpe, B., & Wehrle, K. (2014, August). User-driven privacy enforcement for cloud-based services in the internet of things. In *International Conference on Future Internet of Things and Cloud (FiCloud), 2014* (pp. 191–196). IEEE.

Henze, M., Hermerschmidt, L., Kerpen, D., Häußling, R., Rumpe, B., & Wehrle, K. (2016). A comprehensive approach to privacy in the cloud-based Internet of Things. *Future Generation Computer Systems, 56*, 701–718.

Holler, J., Tsiatsis, V., Mulligan, C., Karnouskos, S., & Boyle, D. (2014). *From machine-to-machine to the internet of things: Introduction to a new age of intelligence.* Academic Press.

Jeschke, S., Brecher, C., Meisen, T., Özdemir, D., & Eschert, T. (2017). Industrial internet of things and cyber manufacturing systems. In *Industrial internet of things* (pp. 3–19). Springer International Publishing.

Kang, R., Dabbish, L., Fruchter, N., & Kiesler, S. (2015, July). My data just goes everywhere: User mental models of the internet and implications for privacy and security. In *Symposium on Usable Privacy and Security (SOUPS)* (pp. 39–52). Berkeley, CA: USENIX Association.

Kravitz, D. W., & Cooper, J. (2017, June). Securing user identity and transactions symbiotically: IoT meets blockchain. In *Global Internet of Things Summit (GIoTS), 2017* (pp. 1–6). IEEE.

Kumar, J. S., & Patel, D. R. (2014). A survey on internet of things: Security and privacy issues. *International Journal of Computer Applications, 90*(11).

Leenes, R., van Brakel, R., Gutwirth, S., & De Hert, P. (2017). Data protection and privacy: (In) visibilities and infrastructures. *Issues in Privacy and Data Protection, 36*(1).

Leminen, S., Rajahonka, M., Westerlund, M., & Siuruainen, R. (2015). Ecosystem business models for the internet of things. *Internet of Things Finland*, 10–13.

Lin, J., Shen, Z., & Miao, C. (2017, July). Using blockchain technology to build trust in sharing lorawan IoT. In *Proceedings of the 2nd International Conference on Crowd Science and Engineering* (pp. 38–43). ACM.

Manu, A. (2016). *Value creation and the internet of things: How the behavior economy will shape the 4th industrial revolution*. Routledge.

Mcardle, E. (2016, May). The new age of surveillance. *Harvard Law Today*. Retrieved from https://today.law.harvard.edu/feature/new-age-surveillance/

Millard, C., Hon, W. K., & Singh, J. (2017, April). Internet of things ecosystems: Unpacking legal relationships and liabilities. In *IEEE International Conference on Cloud Engineering (IC2E), 2017* (pp. 286–291). IEEE.

O'Connor, Y., Rowan, W., Lynch, L., & Heavin, C. (2017). Privacy by design: Informed consent and internet of things for smart health. *Procedia Computer Science, 113*, 653–658.

Pagallo, U., Durante, M., & Monteleone, S. (2017). What is new with the internet of things in privacy and data protection? Four legal challenges on sharing and control in IoT. In *Data protection and privacy: (In) visibilities and infrastructures* (pp. 59–78). Springer International Publishing.

Palattella, M. R., Dohler, M., Grieco, A., Rizzo, G., Torsner, J., Engel, T., & Ladid, L. (2016). Internet of things in the 5G era: Enablers, architecture, and business models. *IEEE Journal on Selected Areas in Communications, 34*(3), 510–527.

Pasquier, T., Singh, J., Powles, J., Eyers, D., Seltzer, M., & Bacon, J. (2017). Data provenance to audit compliance with privacy policy in the internet of things. *Personal and Ubiquitous Computing, (22)*2, 333–344.

Peppet, S. R. (2014). Regulating the internet of things: First steps toward managing discrimination, privacy, security and consent. *Texas Law Review, 93*, 85.

Perera, C., Ranjan, R., Wang, L., Khan, S. U., & Zomaya, A. Y. (2015). Big data privacy in the internet of things era. *IT Professional, 17*(3), 32–39.

Porambage, P., Ylianttila, M., Schmitt, C., Kumar, P., Gurtov, A., & Vasilakos, A. V. (2016). The quest for privacy in the internet of things. *IEEE Cloud Computing, 3*(2), 36–45.

Qin, Y., Sheng, Q. Z., Falkner, N. J., Dustdar, S., Wang, H., & Vasilakos, A. V. (2016). When things matter: A survey on data-centric internet of things. *Journal of Network and Computer Applications, 64*, 137–153.

Richardson, M., Bosua, R., Clark, K., Webb, J., Ahmad, A., & Maynard, S. (2017). Towards responsive regulation of the internet of things: Australian perspectives. *Internet Policy Review, 6*(1).

Sadeghi, A. R., Wachsmann, C., & Waidner, M. (2015, June). Security and privacy challenges in industrial internet of things. In *52nd Design Automation Conference (DAC), 2015* (pp. 1–6). IEEE.

Samaniego, M., & Deters, R. (2016, December). Hosting virtual IoT resources on edge-hosts with blockchain. In *IEEE International Conference on Computer and Information Technology (CIT), 2016* (pp. 116–119). IEEE.

Schneier B. (2016, November). Regulation of the internet of things. Retrieved from https://www.schneier.com/blog/archives/2016/11/regulation_of_t.html

Sharma, P. K., Chen, M. Y., & Park, J. H. (2018). A software defined fog node based distributed blockchain cloud architecture for IoT. *IEEE Access, 6*, 115–124.

Shin, D. H., Shin, D. H., Jin Park, Y., & Jin Park, Y. (2017). Understanding the internet of things ecosystem: Multi-level analysis of users, society, and ecology. *Digital Policy, Regulation and Governance, 19*(1), 77–100.

Sicari, S., Rizzardi, A., Grieco, L. A., & Coen-Porisini, A. (2015). Security, privacy and trust in internet of things: The road ahead. *Computer Networks, 76*, 146–164.

Singh, J., Pasquier, T., Bacon, J., Powles, J., Diaconu, R., & Eyers, D. (2016, November). Big ideas paper: Policy-driven middleware for a legally-compliant Internet of Things. In *Proceedings of the 17th International Middleware Conference* (p. 13). ACM.

Storr, C., & Storr, P. (2017). Internet of things: Right to data from a European perspective. In *New technology, big data and the law* (pp. 65–96). Springer, Singapore.

Svetinovic, D. (2017, April). Blockchain engineering for the internet of things: Systems security perspective. In *Proceedings of the 3rd ACM International Workshop on IoT Privacy, Trust, and Security* (pp. 1–1). ACM.

Thierer, A. D. (2015). The internet of things and wearable technology: Addressing privacy and security concerns without derailing innovation. *Richmond Journal of Law and Technology, 21*, 1.

Turgut, D., & Boloni, L. (2017). Value of information and cost of privacy in the internet of things. *IEEE Communications Magazine, 55*(9), 62–66.

Urquhart, L., Sailaja, N., & McAuley, D. (2017). Realising the right to data portability for the domestic internet of things. *Personal and Ubiquitous Computing, 22*(2), 1–16.

Wang, X., Chen, F., Ye, H., Yang, J., Zhu, J., Zhang, Z., & Huang, Y. (2017). Data transmission and access protection of community medical internet of things. *Journal of Sensors, 2017*.

Weber, R. H. (2015). Internet of things: Privacy issues revisited. *Computer Law & Security Review, 31*(5), 618–627.

Weinberg, B. D., Milne, G. R., Andonova, Y. G., & Hajjat, F. M. (2015). Internet of Things: Convenience vs. privacy and secrecy. *Business Horizons, 58*(6), 615–624.

Xie, M., Huang, M., Bai, Y., & Hu, Z. (2017). The anonymization protection algorithm based on fuzzy clustering for the ego of data in the internet of things. *Journal of Electrical and Computer Engineering, 2017*.

Yan, Z., Zhang, P., & Vasilakos, A. V. (2014). A survey on trust management for internet of things. *Journal of Network and Computer Applications, 42*, 120–134.

Zhang, Y., & Wen, J. (2017). The IoT electric business model: Using blockchain technology for the internet of things. *Peer-to-Peer Networking and Applications, 10*(4), 983–994.

Ziegeldorf, J. H., Morchon, O. G., & Wehrle, K. (2014). Privacy in the internet of things: Threats and challenges. *Security and Communication Networks, 7*(12), 2728–2742.

Chapter 11

IoT Cyber Security—A Discourse on the Human Dimension

The key to growth is the introduction of higher dimensions of consciousness into our awareness.

Lao Tzu

After reading this chapter you will be able to:

- Understand the human dimension of cyber security for a trustworthy IoT environment
- Understand the necessity of a stakeholder approach for IoT cyber security
- Gain an insight on the influence of digital footprints on IoT cyber security
- Understand the importance of collective intelligence for safe and secure IoT systems.

Introduction

Eminent information security expert Bruce Schneier had commented that "Security is only as good as its weakest link, and people are the weakest link in the chain."* This is a very apt comment on end users' role in cyber security, and is

---

* Schneier, B. (2011). Secrets and lies: digital security in a networked world. John Wiley & Sons.

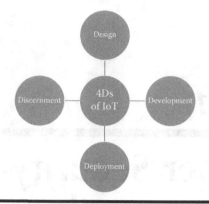

Figure 11.1　Four Ds of IoT technology implementation.

very significant for IoT technology. Although Schneier made this comment about twenty years ago, with the passage of time we are seeing a growing trend in cyber attacks, both internal or external and accidental as well as malicious in nature, across the globe with no sign of subsiding. As a result, Schneier's comment is gaining more relevance with every passing day. Security and privacy are key enablers as well as sacrosanct for realizing the IoT's potential. The human dimension of cyber security in the IoT is of significant importance in all the four stages of IoT implementation: design, development, deployment and discernment.

IoT Cyber Security—A Top-Down Approach?

If we look at the history of computers, we will see that computing efforts were started in order to assist humans in repetitive manual computations through iterative procedures.* Gradually, various activities and business processes have been mapped logically into computing procedures. The process of computing requires the involvement of humans in all stages of development: requirement analysis, algorithm design, programming, data structuring, developing data relations, performance testing and deployment for use. However, in various stages of the advancement of computing, when it has been used as an enabler or as an integral part of business, we have not embedded efficient security features by default into the design of computing. Rather, we have always followed a top-down approach whereby, based on security incidents and the wisdom of experienced people, we have inserted security features in the IT service-delivery framework. We have embraced various methods in this approach for setting up security features in IT services—some of these are based on 'closed-door' policies, whereby no one is provided access to any of the computing features

---

* Campbell-Kelly, M., Aspray, W., Snowman, D. P., McKay, S. R., & Christian, W. (1997). Computer: A history of the information machine. *Computers in Physics, 11*(3), 256–257.

without proper authorization, and others are based on 'open-door' policies, whereby all authentic users are provided authority on all usable features of the software unless there is an approved justification for cases of exclusivity.

With the passage of time, as software has become prone to unauthorized access, we have introduced features of authentication, authorization and accounting for continuous monitoring. As more untoward cyber incidents have started being reported, we have moved to multi-factor authentication and biometric authentication. But has this top-down approach really helped us? If it has helped in a true sense, then there would not be so many incidents of breaches across the globe affecting business as well as normal human life. This question requires introspection to find the right answer. Also, would a bottom-up approach to security that is primarily based on the premise of thinking about and designing security right from the design phase of a particular IT service or product be a better option?

Any approach to IoT cyber security, be it top-down or bottom-up or a mix of both, will not be fully satisfying if human roles and responsibilities are not performed with adequate behaviors conforming to the objective of computing. This failure to solve the issue of information-systems security has often been traced to the engineering principles upon which computer science is based,* while others have attributed it to the insufficient skills of the technologists involved, corporate greed and lack of integrity. It has been argued† that a high degree of ignorance and paranoia about the issue of information-systems security exists on both ends—the provider and the user—and that this is the main cause of the failure of information-systems security.

A Stakeholder Approach to IoT Cyber Security

If we look at stakeholder salience‡ in cyber-security scenarios in the present day, we see that the users of IT services or products have the most salience due to a variety of factors like necessity of use, intention of use and benefit of use, as shown in Figure 11.2 below.

If the necessity and intention of the user are aligned with the design of the software product or service, then the benefit derived out of it will be in sync with the intent with which the software product or service was made. However, if the user's necessity and intention do not map with that of the software product or service, then the benefit derived by the user does not match the intent with which the software product or service was made available. This is where the problem arises with

---

* Flink, C. W. (2002). Weakest link in information system security. Retrieved from http://www
.acsac.org/waepssd
† Ibid.
‡ Mitchell, R. K., Agle, B. R., & Wood, D. J. (1997). Toward a theory of stakeholder identification and salience: Defining the principle of who and what really counts. *Academy of Management Review, 22*(4), 853–886.

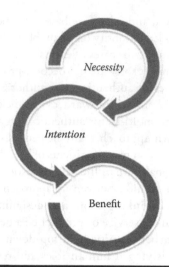

Figure 11.2 The interconnection of human necessity and intention of cyber activities with derived benefits.

respect to safe usage of a software product or service, which can lead to security breaches. This cyber issue ultimately boils down to the human role and responsibility for ensuring cyber security, and human responsibility for cyber security in this case can be broadly partitioned into two sets—those who design, develop and maintain the software product or service and those who use that software product or service as per their necessity and intention of usage. The same perspective holds for IoT scenarios as well. We can encounter security vulnerabilities in an IoT product or service for any or all of the following reasons:

- If it is not properly designed, developed or tested before use
- If there is no proper epistemic guidance for appropriate usage of the smart product or service
- If improper usage is not prevented with punishment or penalty and appropriate resilience measures.

To define appropriate IoT product or service design and usage, we need an intervention with the key stakeholders, including technologists, smart-product or service owners or custodians, regulators, auditors, and governments. The primary responsibility of these stakeholders is to ensure appropriate cyber behavior toward the four Ds of IoT technology (design, development, deployment and discernment) and to provide specific guidelines for assuring this appropriateness.

The OECD recommends that "governments, public and private organizations, as well as individuals share responsibility, based on their roles and the context, for managing digital security risk and for protecting the digital environment; and that co-operation is essential at domestic, regional and international levels."* It emphasizes that information-systems security is a collective responsibility at various strata of life and hence requires due diligence at various layers of society, businesses and governance. Also, the emphasis is on international cooperation, as cyber security is no more a regional or country-level issue. With the advent of the IoT, it will be more pervasive in nature and the effects of IoT security breaches will impact critical smart services in ways that ripple across the globe.

IoT Security and Digital Footprint

IoT security is deeply interlinked with the digital footprints of humans being generated at every moment across the globe from Internet-connected devices. But how aware we are of our own digital footprints is a question that is not easy to answer due to multiple factors. For example, when we access specific websites on the Internet on a daily basis, we are also directed or misdirected to multiple other web pages through links, advertisements or due to the human desire to explore the unknown. However, the user accessing the Internet is not always at fault for their so-called misadventures in exploring in the digital world. It also happens that the websites users view collect information about them with or without consent. Cookies may be installed on the computer during browsing, explicitly or implicitly. In such situations, who should be responsible for providing a clean Internet platform? This primarily depends on the mutual trust and responsibility of the provider of the service or product as well as the user.[†] The onus for ensuring a safe cyber environment also lies on other key stakeholders, like governments, legal and regulatory bodies, auditors and businesses.

With the advent of IoT technology, the cyber world is moving into a completely see-through environment, whereby data about almost all things on earth can be generated round the clock and analyzed to provide unprecedented digital services. The tremendous potential of IoT technology can change the lives of humans in almost all aspects, but it also brings to the fore the concern of compromised security and privacy in the digitally connected life of the near future. The IoT will create a sea of digital footprints for every human on earth, and this will expose the personally identifiable information and behavioral characteristics of humans to other humans to be used for malicious activities.

In a smart city, humans with smart-device connectivity will be integral components of the city cyber systems, and they will operate and provide inputs to these

---

* OECD. (2015). The OECD recommendation on digital security risk management for economic and social prosperity. Retrieved from http://www.oecd.org/sti/ieconomy/digital-security-risk-management.pdf

† Schmidt, J. (2011). Humans: The weakest link in information security. Retrieved from http://www.forbes.com/sites/ciocentral/2011/11/03/humans-the-weakest-link-in-information-security/

interdependent systems and receive outputs from these systems. These citizens can create security vulnerabilities in the cyber system unknowingly or as malicious insiders, which can lead to cyber attacks and cyber disasters. Considering humans as the live components of smart interdependent systems in smart cities, we can also think of them as vulnerable intrusion points if a minimal level of standardized cyber awareness and usage behavior are not maintained.

The Importance of Collective Intelligence for a Secured and Trustworthy World of 'Things'

Cities getting smarter with IoT technology need effective cyber policies that aim to enhance the epistemic sense of citizens for the judicious use of this potential technology in order to prevent oneself and others from getting trapped in social engineering attacks and from committing casual digital mistakes that can make smart systems vulnerable to cyber attacks. Smart cities need a network of sensors and a network of 'Thing'-aware citizens. We have to build a state of mindfulness in citizens for appropriate cyber behavior in smart cities, whereby every digital action they perform has security awareness in mind.

Governments and city councils can organize cyber-behavior change campaigns for individuals and organizations to develop good cyber security habits. The 'Cyber Aware' initiative of the UK Government is one such initiative, which "aims to drive behaviour change amongst small businesses and individuals, so that they adopt simple secure online behaviours to help protect themselves from cyber criminals."*

Cities like Boston, Buenos Aires and Santander are utilizing the collective intelligence of citizens for smart-service improvement and better decision making. "Citizens equipped with mobile phones capable of capturing, transmitting, and receiving information form a digital sidewalk ballet, contributing localized bits of knowledge, ideas, and opinions that lead to smarter decisions."† The 'Clean India Mission'‡ which covers over 4000 towns, has a mobile app for citizens to photograph and report streets in need of cleaning or with problems that need to be fixed by the authorities.§ Similarly, we can utilize citizens' collective intelligence to develop a cyber awareness of society and to ensure a rational cyber behavior with smart systems. Untoward cyber incidents in smart-city systems can be reported and

---

* Cyber Aware. (n.d.). Retrieved from https://www.cyberaware.gov.uk/about-us
† Eggers, W. D., Guszcza J., & Greene, M. (2017). Making cities smarter. *Deloitte Review,* *20*. Retrieved from https://www2.deloitte.com/content/dam/Deloitte/de/Documents/human -capital/DR20_Making_cities_smarter.pdf
‡ Clean India Mission. (n.d.). Retrieved from https://swachhbharat.mygov.in/
§ Engasser F., & Saunders T. (2015). Role of citizens in India's smart cities challenge. Retrieved from http://www.worldpolicy.org/blog/2015/11/03/role-citizens-india%E2%80%99s-smart-cities -challenge

voluntarily acted upon by citizens through apps. This can help develop a techno-epistemic society for preventing the spread of cyber attacks on interdependent smart-city systems by quickly locking down affected systems and through collective recovery initiatives.

An overarching body is required to evaluate, direct and monitor IoT product and service designs, appropriate user behavior with 'Things' and judicious collection and sharing of personally identifiable information by businesses and agencies, including smart initiatives in the guise of beneficence. As individuals, we also have a responsibility to be cyber-aware and IoT technology-competent for rational cyber behavior. These initiatives from IoT stakeholders will collectively help to eliminate the weak links in the IoT ecosystem and lead to a less vulnerable world of 'Things' for all.

Suggested Reading

Boyes, H. (2013). *Resilience and Cyber Security of Technology in the Built Environment*. Retrieved from http://www.cpni.gov.uk/documents/publications/2013/2013063-resilience _cyber_security_technology_built_environment.pdf?epslanguage=en-gb

BSI. (2014). *PAS 181 smart city framework*. Retrieved from http://www.bsigroup.com/en-GB /smart-cities/Smart-Cities-Standards-and-Publication/PAS-181-smart-cities-framework/

Buchanan B. (2014). In cybersecurity, the weakest link is ... you. *TheConversation.com*. Retrieved from http://theconversation.com/in-cybersecurity-the-weakest-link-is-you -33524

Cabinet Office. (2012). *A summary of the 2012 sector resilience plans*. London, UK: Cabinet Office. Retrieved from https://www.gov.uk/government/uploads/system/uploads /attachment_data/file/271349/Summary-2012-Sector-Resilience-Plans.pdf

Caralli, R. A. et.al. (2007). *Introducing OCTAVE Allegro: Improving the Information Security Risk Assessment Process*. Software Engineering Institute. Retrieved from http:// resources.sei.cmu.edu/asset_files/TechnicalReport/2007_005_001_14885.pdf

Chaudhuri, A. (2011). Enabling effective IT governance: Leveraging ISO/IEC 38500: 2008 and COBIT to achieve business–IT alignment. *EDPACS*, *44*(2), 2–18.

CPNI. (2015). *Good practice guide–Process control and SCADA security*. Retrieved from http://www.cpni.gov.uk/Documents/Publications/2008/2008031-GPG_SCADA _Security_Good_Practice.pdf

Curtis, B. H. (2001). *People capability maturity Model. (2.0)*. Carnegie Mellon Software Engineering Institute. Retrieved from http://cmmiinstitute.com/resources/people -capability-maturity-model-p-cmm

Department for Business Innovation & Skills, UK Government. (2013). *Smart cities: background paper*. Retrieved from https://www.gov.uk/government/uploads/system /uploads/attachment_data/file/246019/bis-13-1209-smart-cities-background-paper -digital.pdf

European Commission. (2015). *The Internet of things*. Retrieved from http://ec.europa.eu /digital-agenda/en/internet-things

Falco, G. J. (2015). City resilience through data analytics: A human-centric approach. *Procedia Engineering*, *118*, 1008–1014.

Flink, C. W. (2002). Weakest link in information system security. *WEAPSSD Proceedings.* Retrieved from http://www.acsac.org/waepssdFreeman,

Gilbert D. (2013). Employees are 'weakest link' in cyber security chain. *International Business Times.* Retrieved from http://www.ibtimes.co.uk/employees-weakest-link -cyber-security-chain-494870

Government Office of Science. (2012). *Infrastructure and Resilience.* London, UK: Foresight. Retrieved from https://www.gov.uk/government/uploads/system/uploads/attachment _data/file/286993/12-1310-infrastructure-and-resilience.pdf

IEEE. (2015). *Smart cities.* Retrieved from http://smartcities.ieee.org/about.htm

Information security: 'only as strong as the weakest link in the chain'. *Knowledge@Wharton* (2013, Feb. 25). Retrieved from http://knowledge.wharton.upenn.edu/article/information -security-only-as-strong-as-the-weakest-link-in-the-chain/

IoT6. (2014). *IPv6 for IoT.* Retrieved from http://iot6.eu/ipv6_for_iot

ISO/ IEC. (2008). *Corporate governance of information technology.* Switzerland: ISO. Retrieved from http://www.iso.org/iso/catalogue_detail?csnumber=51639

IT Governance Institute. (2001). *Board briefing on IT governance* (2nd Edition ed.). Information Systems Audit and Control Association.

Korolov, M. (2015, Apr. 17). Attacks against industrial control systems double. *ITWorld. com.* Retrieved from http://www.itworld.com/article/2911634/attacks-against-industrial -control-systems-double.html

Little, R. G. (2002). Controlling cascading failure: Understanding the vulnerabilities of interconnected infrastructures. *Journal of Urban Technology, 9*(1), 109–123.

Mitchell, S. et. al. (2013). *The internet of everything for cities.* Retrieved from http://www .cisco.com/web/strategy/docs/gov/everything-for-cities.pdf

NCIIPC. (2013). *Guidelines for protection of national critical information infrastructure.* Government of India. (n.d.). Retrieved from http://perry4law.org/cecsrdi/wp-content /uploads/2013/12/Guidelines-For-Protection-Of-National-Critical-Information -Infrastructure.pdf

OECD. (2015). *Digital security risk management for economic and social prosperity: OECD recommendation and companion document.* Retrieved from http://www.oecd.org/sti /ieconomy/digital-security-risk-management.pdf

R. E. (2001). A stakeholder theory of the modern corporation. *Perspectives in Business Ethics Sie, 3,* 144.

Rinaldi, B. P. (2001). Identifying, understanding, and analyzing critical infrastructure interdependencies. *IEEE Control Systems, 21*(6), 11–25.

Schmidt, J. (2011, Nov. 3). Humans: The weakest link in information security. *Forbes.* Retrieved from http://www.forbes.com/sites/ciocentral/2011/11/03/humans-the-weakest -link-in-information-security/

Schneider Electric. (n.d.). Smart cities. Retrieved from http://www2.schneider-electric.com /sites/corporate/en/solutions/sustainable_solutions/smart-cities.page

Schneier, B. (2000). *Secrets and lies.* New York: John Wiley & Sons.

Townsend, A. M. (2013). *Smart cities: Big data, civic hackers, and the quest for a new utopia.* (1st ed.). London: W.W.Norton & Company.

Watson, J. (2014). *The Resilience of City Systems* [PDF document]. Retrieved from http:// www.2014.csdm-asia.net/IMG/pdf/CSDM_Jeremy_Watson.pdf

Appendix 1: The Proactive and Preventive Privacy (3P) Framework for IoT Privacy by Design

Abhik Chaudhuri and Dr. Ann Cavoukian

IDEA IN BRIEF

The Problem

The IoT is providing new services and insights by sensing contextual data, but there are growing concerns of privacy risks from users that need immediate attention.

The Reason

IoT devices and smart services can capture personally identifiable information (PII) without user knowledge or consent. IoT technology has not reached the desired level of maturity to standardize security and privacy requirements.

The Solution

IoT Privacy by Design is a user-centric approach for enabling privacy with security and safety as a 'win-win' positive outcome of IoT offerings, irrespective of business domain. The proactive and preventive privacy (3P) framework proposed in this paper should be adopted by the IoT stakeholders for building end user trust and confidence in IoT devices and smart services.

About This Content

The novel IoT privacy framework discussed here has been published as a paper in EDPACS Journal (Taylor and Francis, US) in February 2018.

Dr. Ann Cavoukian is best known for her creation of Privacy by Design—unanimously adopted as an international framework for privacy and data protection in 2010, now translated into 38 languages. Appointed as the information and privacy commissioner of Ontario, Canada in 1997, Dr. Cavoukian served for an unprecedented three terms as commissioner. As of July 1, 2014, she began a new position at Ryerson University as the executive director of the Privacy and Big Data Institute—Where Big Data meets Big Privacy.

Introduction

The global society is moving rapidly toward a 'connected' future. IoT devices and IoT-enabled smart services are gradually catching up in the societal-usage meter across the globe, with increases in smart devices and smart service offerings in various segments like wearable devices, connected homes, connected cars, connected healthcare, smart-city services and many other potential smart-service offerings. Various studies are predicting a multi-fold growth of IoT devices in billions across the globe over the next five years. However, privacy is a key concern for the smart services across verticals. To address privacy concerns, we need privacy enablers embedded in the design of IoT devices and smart services from the conceptual stage to make them reliable and trustworthy.

The principles of IoT Privacy by Design discussed in this paper are aimed at addressing privacy concerns of IoT devices and smart services. These principles are adaptive extensions of the foundation principles of Privacy by Design and are based on the stakeholder needs in IoT services, and considering the users and smart citizens as key stakeholders, the protection of their personally identifiable information (PII) is the primary focus.

IoT Privacy by Design extends the 'trilogy' of encompassing applications—IT systems, accountable business practices, physical design and networked infrastructure—with the fourth dimension of 'sensors.' Anyone using these adaptive principles of IoT Privacy by Design will need to address the requirements of the seven foundation principles as well, because the adaptive principles and foundation principles are complementary to each other.

The proactive and preventive privacy (3P) framework proposed in this paper can be adopted by the IoT stakeholders for building end user trust and confidence in IoT devices and smart services.

Ushering in the Era of Sensors

The IoT can provide a ubiquitously connected world and create huge oceans of data logged in a continuous mode based on a predefined and fine-tuned context. Sensors are playing a critical role in implementing the IoT. Sensors create the layer of perception by sensing contextual data as per defined parameters of the smart service from the environment where these are deployed. The rapidly evolving IoT technology is utilizing this additional dimension of sensors to capture data from contexts that were not previously possible on such a massive scale. The parameters that define the context can be changed by authorized command and control of the actuator. Depending on the smart-service design, a huge number of sensors might be deployed to gather contextual data. The sensors currently in use have limited computing power, making it difficult to render strong encryption during data communication from the sensors to the gateway. Sending back control data to the actuator in encrypted format is also not feasible for the same reason.

The data from the sensors are communicated through wired and wireless networks to be stored, processed and analyzed for providing the context-based smart services. While new business and service models are being designed in different domains using perceived data—which makes IoT technology fascinating—this is also bringing up concerns about user privacy. The common man might find it difficult to understand the architecture of these sensors and what data the IoT sensors are collecting about them, because the details of the captured data are either not shared fully by the device manufacturers and service providers or the users are unaware that they should have an operational know-how of the IoT device and smart service for their personal benefit.

The IoT Ecosystem

The IoT, being an emerging technology, has not yet reached a mature and stable level in the technology adoption life cycle. It is a complex ecosystem with multiple technologies and a large number of players across various layers. The IoT ecosystem comprises various device and sensor manufacturers, chipset vendors, platform vendors, cloud-infrastructure service providers, developer communities, system integrators, value-added service providers, research consortiums, standards bodies and regulatory authorities. These are early days for this emerging technology, but we are already seeing security and privacy concerns being raised for the various IoT products and services that are currently being offered. Ensuring security and privacy for IoT devices and smart services is a necessity that needs immediate attention to realize the IoT's potential.

Privacy a Key Enabler for the IoT

As IoT-enabled smart services become popular, the gradual overlap of IoT devices and services with our daily routines will also result in increased concerns of user privacy if this issue is not addressed now, at the nascent stage of this technology. This is because the capability of IoT devices to talk to each other (M2M communication) and the technical capability to share contextual data between these devices can aid in the profiling of users and gauging user behavior by data aggregation, re-identification and analytic methods for business benefits that might happen without the knowledge or consent of the user. For example, a digital impression of our daily lives can be created by smart devices monitoring our TV-watching preferences, analyzing the energy usage from various appliances in our home, tracking the location of our cars, monitoring our health parameters or tracking our food preferences by communicating with the smart refrigerator in our home. The spatial and temporal aspects of IoT-enabled smart services can enable data aggregation that can help businesses infer various traits about the users of these smart services.

IoT-enabled service offerings can be interconnected and interdependent, as in smart cities, where home security systems can feed data and alerts to the city surveillance system maintained by the city police or government, smart refrigerators can order food from online grocery stores on behalf of their owners, or smart health monitoring systems can send health data from smart, wearable health-monitoring devices to health insurers.

IoT service interfaces can make contextual data usage opaque to the common man, who may leave IoT devices on default settings without any fine-tuning for privacy conservation. Also, not all consumers or users read the fine print of IoT device-usage policies, and not all of them understand the potential privacy breaches that can result from accepting the default usage settings of these devices. The minor negligence or inabilities of users can lead to misuse of PII and privacy invasion, ultimately affecting the trust and reputation of these smart devices and services. So privacy should be considered as a key enabler in realizing the potential of IoT technology, in tandem with other key enablers like security, safety, stakeholder requirements, regulatory requirements and public policies. These key enablers are the foundation for a trustworthy, connected world of 'Things,' as depicted in Figure A1.1.

The IoT Data Components and Privacy Concerns

The IoT is providing new dimensions in business models and service offerings across industries. For example, smart-home applications are gaining popularity among citizens across the globe. Various offerings are available in the market for room-ambience control, security and surveillance and appliance-based smart offerings like smart refrigerators and smart TVs.

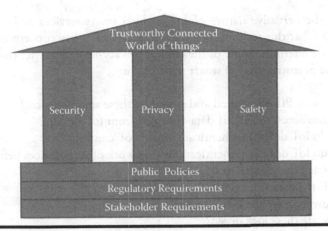

Figure A1.1   Privacy as a key enabler for a trustworthy, connected world.

For the healthcare industry we are seeing a surge in various offerings that are either available in the market or are being prototyped. For example, fitness-related smart, wearable devices are already quite popular, and we have various kinds of service offerings in this segment from multiple well-known as well as lesser-known product companies. Futuristic smart-healthcare applications like IoT-enabled remote monitoring devices for diabetics, heart and pacemaker-monitoring devices, monitoring devices for patients with kidney ailments and health monitors for senior citizens are already in various stages of realization from different product and service vendors.

The IoT is also promising to rapidly evolve the retail industry with new smart-service offerings like smart billing and payment systems eliminating the point of sales, smart fitting rooms and agile display of product displays on shelves based on customer gestures.

Although there can be various means of data modeling and service designs for smart services, the key IoT data components for these offerings can be broadly categorized into five types:

■ Data collected by edge devices like wireless sensors, IP cameras, barcode readers, RFID readers and GPS devices
■ Data at gateway devices flushed periodically from edge devices by wired and wireless networksData sent to the cloud by gateway devices for analytical processing, storage and application-based output
■ API-based data interchange for various smart-service offerings from machines to machines and from machines to users
■ 'Control' data sent back to edge devices and sensors for controlling or fine-tuning the context of data gathering.

Due to the pervasive nature of IoT-enabled smart services and the limitless opportunities that this technology provides, privacy is a key concern for the users of these smart offerings. Some of the critical privacy-related queries that are being raised by the potential users of smart services are:

- How much PII is captured and stored by these smart services?
- Who has access to the PII data obtained from IoT devices?
- How do IoT devices authenticate senders of 'control' data?
- How do IoT devices authenticate in a network of such devices before sending contextual data for smart offerings?
- How is the data from the IoT sensors logged?
- For how long are these logs stored?
- Where are these logs stored?
- Who has access to these logs?
- Is the data from IoT devices copied to multiple locations across geographies for backup?
- Is the data safe in transit?
- Is any customer's PII being compromised by any means?
- Who owns the PII data collected from IoT devices?
- Does the contextual data gathered by IoT devices remain anonymous in the process?
- Can smart service providers exploit or share PII data for business benefits without user consent?
- In the IoT service chain, how are the user's privacy requirements mapped to the various components of end-to-end IoT service?
- Are IoT devices monitoring more parameters than have been disclosed, advertised or mentioned in the user agreement?

As we move into the digital era of 'Things,' we have to provide satisfactory answers to the above queries to provide reliable and trustworthy smart services that will be compliant with regional and global privacy requirements from all stakeholders, including legal and regulatory authorities.

From the IT era we have learned from our mistakes in not considering security and privacy requirements early in the product and service design phase. The penalty that we are paying for this is enormous in terms of reputation, trust and cost. We should not repeat the same mistakes with the IoT. This is the right opportunity to learn from history and consider privacy along with security in the IoT products and service that are being conceived. The principles of IoT Privacy by Design discussed here will help to address the privacy concerns for reliable and trustworthy smart products and services.

The Seven Adaptive Principles of IoT Privacy by Design

IoT privacy needs should be addressed to realize the potential of this technology. A Privacy-by-Design approach is required for these emerging technologies and smart services to address growing privacy concerns and needs. Below we present the seven principles of IoT Privacy by Design. These are adaptive principles based on Dr. Cavoukian's Privacy by Design principles that have been accepted globally.

1. Proactively Prevent Privacy-Invasive IoT Events

Privacy-enhancing capabilities should be built into IoT devices and smart services with an aim to prevent privacy-invasive events. Anticipating IoT privacy events early during the ideation and development phases will help to prevent reactive responses to privacy breaches that can cause distrust among users. Satisfactory answers to the following questions are required to take a proactive approach:

— Does the IoT device or smart service collect and use any PII?
— Will the user's PII be shared with other IoT service providers and third parties?
— If the answer to the above questions is 'Yes', then is there any provision to inform the user and record his/her consent for the collection and use of PII?
— Is there any kind of privacy vulnerability in the IoT device or service?
— What are the privacy regulations that require compliance for the IoT device or smart service? What provisions have been considered to meet the compliance requirements?

The IoT device manufacturers and smart-service providers can follow the **"Proactive and Preventive Privacy (3P) Framework"** discussed in the next section to proactively prevent privacy-invasive events in their devices and services.

2. Ensure IoT Privacy by Default

PII data should be protected by default settings built into IoT devices and smart services, with no additional individual effort necessary to protect personal data. The accountability of privacy preservation should be on the IoT device manufacturers and smart-service providers. This will help to build trustworthy IoT offerings for wider acceptability and adoption. We have to understand that the technology world is not an even playing-field for end users across the globe due to variations in knowledge, awareness and human capabilities in adopting new technologies. A default privacy setting will automatically protect PII without the need of user intervention.

3. Embed Privacy-Enhancing Capabilities into IoT Service Design and Device Architecture

Privacy should be considered integral to IoT device and service architecture. Due to the ubiquitous nature of IoT technology, the risk of privacy breaches is multidimensional in the form of identity privacy, location privacy, digital-footprint privacy and user profiling by data aggregation from search queries. By identifying sensitive data components early in the design phase and embedding privacy-enhancing capabilities into IoT device architecture and smart-service design, we can have reliable and trustworthy IoT offerings that comply with privacy requirements without affecting core functionality.

4. Adopt a Stakeholder Approach to IoT Privacy for Full Functionality, Positive Sum Outcome

The IoT ecosystem has multiple stakeholders, as mentioned earlier, who play significant roles in providing end-to-end IoT service. End users are also key stakeholders in this ecosystem. The privacy warp should run through the fabric of IoT components as a key enabler for all stakeholders to provide full functionality, along with other key requirements like security and safety. Privacy with security and safety results in a positive sum outcome for IoT offerings for all stakeholders, as there are no compromising factors like 'privacy at the cost of security/safety' or vice versa.

5. Provide Full Lifecycle Protection of IoT Data for End-to-End Security and Privacy

The contextual data collected by IoT devices and smart services should be preserved with appropriate security and privacy measures for the entire duration of the data lifecycle and should then be destroyed, ensuring that there is no remanence. This will prevent IoT data aggregation and inference-based PII construction for malicious intents and privacy breaches.

6. Opt for a Verification-Based Trust Approach to IoT

A verification-based trust approach to IoT technology, devices and data components is necessary for transparency in IoT operations. Provisions for the independent verification of IoT operations provide visibility and assurance to all stakeholders that the IoT functions are operating as per the stated objectives.

7. Consider Users at the Core of IoT Services

The IoT is a promising technology that will provide new service offerings to end users in various facets of life, as mentioned earlier (not exhaustive). End-user requirements and interests should be considered at the core of IoT device and service designs. Privacy being one of the key user requirements, a 'users first' strategy for the IoT will help build user trust and confidence in these IoT offerings and ensure wider acceptance.

Proactive and Preventive Privacy (3P) Framework

To ensure proactive and preventive privacy, we can follow the eight-step 3P framework, depicted below in Figure A1.2, for all digital technologies including the IoT.

The IoT is a near real-time experience that needs continuous monitoring of its data and service components to ensure privacy capabilities are appropriate. Applying this 3P framework to design IoT-enabled smart services eliminates reactive-response scenarios to privacy breaches. The following eight phases of this privacy framework consider the privacy of IoT devices and services as their objective:

1. **'Define' IoT Service Design and Operation Blueprint**
   IoT offerings should be defined with a service design having specified objectives of data collection and usage. All IoT components in the operation blueprint should meet these objectives, as data flows from source to sink.
2. **'Develop' IoT Data Flows, Application Interfaces, Infrastructure and Network Layouts Based on Stakeholder Needs**
   IoT data flows should be clearly developed, considering stakeholder needs across application interfaces, infrastructure and network layouts so that data collection, data transmission and processing meet privacy requirements.

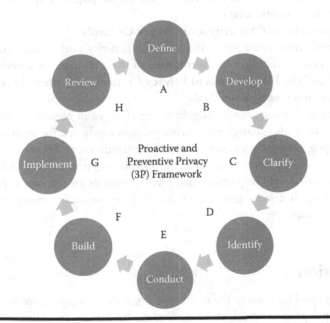

Figure A1.2   Proactive and Preventive Privacy (3P) Framework.

3. **'Clarify,' Document and Limit Purposes for Collecting and Using Personal Data**

   If any collection and usage of personal data is an integral part of an IoT service, then it should be clarified and documented for the knowledge of all stakeholders. The purpose should be limited to meet the stated objective of the IoT service to minimize privacy-invasive events and to eliminate stakeholder concerns.

4. **'Identify' All Security and Privacy Risks**

   Identify all IoT data-security and privacy risks at the data-component level from all sources throughout the data life-cycle.

5. **'Conduct' a Privacy-Impact Assessment of all IoT Devices and Data Components**

   A privacy-impact assessment of IoT devices and data components will be beneficial to identify the collection and usage of PII data. It will also help IoT device and service providers design effective processes to handle personal data and mitigate privacy risks.

6. **'Build' IoT Privacy-Enhancing Capabilities**

   Build privacy-enhancing capabilities into the IoT service design by appropriate technical methods like anonymization, de-identification, metadata encapsulation and other relevant means on the physical, operational, contextual and semantic layers.

7. **'Implement' IoT Security and Privacy Controls**

   Implement operational and systemic security and privacy controls with appropriate validation and verification of these controls on a periodic basis.

8. **'Review' the Effectiveness of Privacy Controls and Identify New Privacy Risks on a Continuous Basis**

   IoT technology is evolving, bringing changes in requirements and priorities. This is also changing the privacy-threat landscape for smart services and bringing up new regulatory-compliance requirements. Reviewing the effectiveness of IoT privacy controls and identifying new privacy risks on a continuous basis will help ensure that privacy controls are relevant as per the need.

   An application—use-case of the 3P Framework for smart homes is discussed below.

Conclusion

With the gradual increase in the number of smart devices connected to the Internet and the rapid pace at which these devices are becoming an intimate part of our daily life, there is also an increase in threat to our privacy. IoT Privacy by Design is an absolute essential today to ensure that we do not fall behind in ensuring data privacy for the multifaceted applications of the emerging IoT technology that promises a connected world of all things. The IoT Privacy by Design principles and the Proactive and Preventive Privacy (3P) Framework for IoT discussed here guide us to bake privacy measures into IoT device architectures and smart service designs with transparency and user centricity.

3P Framework Application Use-Case: Smart Home

The Scenario

A Smart Home solution that has the following IoT components:

- Smart appliances like a smart TV and a smart refrigerator
- A smart ambience-control system for monitoring and controlling room temperature, humidity, light intensity and air quality
- A smart energy-management system for triggering electrical appliances like washing machines and dish washers in off-peak hours and for gauging appliance-specific usage and energy consumption for analytics-based usage recommendations for reduced billing.

The Challenge

PII gathered, analyzed and inferred by IoT device manufacturers, IoT application service providers and other third parties without the data subject's cognizance and consent.

The Solution

An eight-step proactive and preventive privacy approach for the smart-home solution using the 3P framework, discussed below.

Step 1: *'Define' the smart-home service design and operation blueprint*
Define the objectives of data collection and usage for the smart-home solution. The necessity of data should be solely for providing the smart service. For example, the smart refrigerator should not gather personal data to be sent to the manufacturer that has information about the kind of food being bought without user consent. Similarly, the smart TV's capabilities for recording in-room personal conversations or behavioral visuals should be disabled by default. The service design should clearly define what data is collected, stored and processed with a specific process to erase these data after addressing the defined needs or based on user requirements. The questions that have to be addressed in this step are:

- What is the objective of this smart-home solution from a smart-service perspective?
- Who are the stakeholders in this smart-home solution?

- Which smart devices will be integrated with this solution?
- What user data is required to enable this smart service?
- Which smart devices are feeding data to other smart devices in this service design?
- What are the data-collecting components, data-storing components and data-processing components in this smart-service operation blueprint?
- What is the authorization and authentication mechanism for these smart devices?
- What are the communication modes (wired, wireless, encrypted or unencrypted) between these smart devices?
- Are the data stored at the edges or in the cloud?
- Are the contextual data with PII content gathered with user consent?
- What security features have been enabled for the data that is collected, stored and in process?
- How is the data life-cycle designed for this smart service?

Step 2: *'Develop' the IoT data flows, application interfaces, infrastructure and network layouts based on stakeholder needs*
The key stakeholders for the smart-home service include the users at the core, the smart service provider with other third-party service providers, the smart city council, device manufacturers, lawmakers, regulators, auditors and others.

Each of these stakeholders will have their own specific requirements. However, keeping the end user's requirement at the core, the smart-home solution should be built in such a manner that the data flow from the sensors, the data interchange between the smart-service application interfaces and the data residing in the infrastructure components like gateways and networked storages are all predefined in the architecture design in compliance with legal and regulatory requirements. Wherever there are no legal or regulatory compliance requirements, the smart-home solution should be tuned to default usage parameters, meaning muted capabilities for PII collection.

For example, the data-flow network can be restricted to sink the data in a localized storage for a smart-home service if the end user is not willing to send the data to a third-party cloud. The user might not be comfortable with their smart refrigerator talking to other integrated smart devices like a smart TV. To cater to such needs, the extended capabilities of digital smartness should be muted by default with options to enable with user consent.

Step 3: *'Clarify,' document and limit purposes for collecting and using personal data*
The default design of the metadata on the smart devices in a smart home should be such that these devices are not collecting any personal data. For example, the smart energy-management system of a smart-home solution should not gather data on the usage behavior of the washing machine and the specific details of the types of clothes washed. Similarly, the smart refrigerator should not log details of medicines

and vaccines that are refrigerated. However, if the smart ambience-control system requires gathering usage data to determine the room-comfort index for the user, then it should be done with clearly documented instructions and capabilities that can be enabled with the user's consent.

Step 4: *'Identify' all security and privacy risks*
Perform security and privacy risk audits periodically on the data components captured from all sources defined in the smart-home service design. This will help the smart-home service provider identify security and privacy vulnerabilities and take necessary action.
Security risks should be considered from the following perspectives:

- The security of the data interchange from the smart devices to gateways and from gateways to storage devices
- The physical security of the smart devices
- The identity and access management of the smart devices
- The end-to-end control of devices, data and networks.

The privacy risks that should be addressed are:

- User-identity privacy
- User-location privacy
- Digital-footprint privacy
- Personal-behavior privacy
- Health-data privacy.

Step 5: *'Conduct' a PIA of all IoT devices and data components*
Perform a PIA based on the identified risks from the smart devices, data components and user needs by addressing the question "Is the user's privacy impacted if this risk is not mitigated?"

For an affirmative answer to the above question for all identified risks, revisiting the smart-service design might be necessary to address any privacy gaps. It might also require disabling specific usage features of the smart-home solution if it is using the user's PII without consent.

Step 6: *'Build' IoT privacy capabilities*
Based on the findings from the PIA, utilize various technical means to enhance the privacy capabilities of the smart-home solution. For example, anonymize the contextual data from the smart refrigerator that contains the user's PII, like name and location, before processing. The smart ambience-control system does not need to collect PII to monitor room-ambience parameters, so a metadata design that considers collecting these along with the functional data can be encapsulated to prevent unauthorized access. The smart energy-management

system does not need any PII for operation, so utilize de-identification techniques to hide PII components from the data gathered by the sensors for this smart service.

Step 7: *'Implement' IoT security and privacy controls*
A standard operational-control procedure should be documented by the smart-home service provider for ready reference by technicians and data-processing teams with mention of specific security and privacy-control steps for the smart service. A documented self-help checklist for the security and privacy of the smart home with emergency help-desk contacts can be shared with users to help them check and fine tune the default settings on their smart devices as per need. Periodic internal security and privacy audits by the smart-home service provider can help them identify the control gaps that need immediate attention for risk mitigation.

Step 8: *'Review' the effectiveness of privacy controls and identify new privacy risks on a continuous basis*
A continuous privacy-capability improvement plan should be developed by the smart-home service provider to review the effectiveness of existing controls and to address legal and regulatory compliance requirements and user needs. The ultimate aim of the smart-home solution should be a trustworthy smart service that adheres to the user's privacy and also complies with legal and regulatory requirements.

Suggested Reading

Baldini, G., Peirce, T., Botterman, M., Talacchini, M. C., Pereira, A., & Handte, M. (2015). IoT governance, privacy and security issues. *Position paper, European Research Cluster on the Internet of Things.*

Brodsky, I. (2016). The race to create smart homes is on. Retrieved from http://www .computerworld.com/article/3062002/home-tech/the-race-to-create-smart-homes -is-on.html

Cavoukian, A. (2006). The case for privacy-embedded laws of identity in the digital age. Retrieved from https://www.ipc.on.ca/images/resources/up-7laws_whitepaper.pdf (accessed on 06/11/2015).

Cavoukian, A. (2008). Privacy in the clouds. *Identity in the Information Society, 1*(1), 89–108. Retrieved from https://www.ipc.on.ca/wp-content/uploads/2008/05/priva cyintheclouds.pdf

Cavoukian, A. (2011). The 7 foundational principles: Implementation and mapping of fair information practices. *Information and Privacy Commissioner's Office, Ontario, Canada.*

Cavoukian, A. (2012). Privacy by design and the emerging personal data ecosystem. *Privacy By Design.* Retrieved from https://www.ipc.on.ca/wp-content/uploads/Resources/pbd -pde.pdf

Cavoukian, A. (2012). Privacy by design: Origins, meaning, and prospects for ensuring privacy and trust in the information era. Retrieved from www.igi-global.com/chapter /privacy-design-origins-meaning-prospects/61500

Cavoukian, A., & Tapscott, D. (1996). *Who knows: Safeguarding your privacy in a networked world.* McGraw-Hill Professional.

Cavoukian, A., & Hamilton, T. J. (2002). *The Privacy Payoff: How successful businesses build customer trust.* McGraw-Hill Ryerson.

Cavoukian, A., Chibba, M., & Stoianov, A. (2012). Advances in biometric encryption: Taking privacy by design from academic research to deployment. *Review of Policy Research, 29*(1), 37–61.

Cavoukian, A., Fisher, A., Killen, S., & Hoffman, D. A. (2010). Remote home health care technologies: How to ensure privacy? Build it in: Privacy by design. *Identity in the Information Society, 3*(2), 363–378.

Cavoukian, A., Stoianov, A., & Carter, F. (2008). Keynote paper: Biometric encryption: Technology for strong authentication, security and privacy. In *Policies and research in identity management* (pp. 57–77). Springer, Boston, MA.

Cavoukian, A., Taylor, S., & Abrams, M. E. (2010). Privacy by design: Essential for organizational accountability and strong business practices. *Identity in the Information Society, 3*(2), 405–413.

Chaudhuri, A. (2015). Address security and privacy concerns to fully tap into IoT's potential. White paper. Retrieved from https://www.researchgate.net/publication/283448081 _Address_Security_and_Privacy_Concerns_to_Fully_Tap_into_IoT's_Potential

Courtin, G. (2015). Five ways retailers can start using IoT today. Retrieved from http:// www.zdnet.com/article/five-ways-retailers-iot-today/

Davis J. S. (2016). Nest, other IoT devices, sent user info in the clear. SC Magazine. Retrieved from http://www.scmagazine.com/nest-other-iot-devices-sent-user-info-in -the-clear/article/466616/

Giannoni-Crystal, F. and Haynes Stuart, A. (2016). The internet-of-things (IoT) (or internet of everything)—Privacy and data protection issues in the EU and the US. Retrieved from http://apps.americanbar.org/webupload/commupload/ST230002 /siteofinterest_files/INFORMATION_LAW_JOURNAL-volume7_issue2.pdf

GOV.UK. (2013). Smart cities: Background paper. Retrieved from https://www.gov.uk /government/uploads/system/uploads/attachment_data/file/246019/bis-13-1209 -smart-cities-background-paper-digital.pdf

IEEE. (n.d.). Smart cities. Retrieved from http://smartcities.ieee.org/about.html

ITU. (2012). Overview of the internet of things. Retrieved from http://www.itu.int /rec/T-REC-Y.2060-201206-I

Mobile Working Group. (2015). Security guidance for early adopters of the internet of things (IoT). Retrieved from https://downloads.cloudsecurityalliance.org/whitepa persSecurity_Guidance_for_Early_Adopters_of_the_Internet_of_Things.pdf

Motti, V. G., & Caine, K. (2015, January). Users' privacy concerns about wearables. In *International Conference on Financial Cryptography and Data Security* (pp. 231–244). Springer, Berlin, Heidelberg.

Shoemaker J. (2016). Privacy commissioner targets IoT health devices in sweep. Retrieved from http://www.lexology.com/library/detail.aspx?g=eec63029-2ea2-4ad2-aa16 -b2af935edbbd

Sullivan B. (2016). Data breaches give rise to 'privacy conscious' smart home hubs. Retrieved from http://www.techweekeurope.co.uk/e-regulation/data-breaches-privacy -conscious-smart-home-hub-190295

Thierer, A. D. (2015). The internet of things and wearable technology: Addressing privacy and security concerns without derailing innovation. Retrieved from http://dx.doi .org/10.2139/ssrn.2494382

Voas, J. (2016). Primitives and elements of internet of things (IoT) trustworthiness. NIST IR 8063, *Nat' l Inst. Standards and Technology.*

Appendix 2: Information Risk Assessment Frameworks and Standards for Digital Services

Introduction

The relevant information risk-assessment frameworks and standards for digital services have been provided here for ready reference. The information has been divided into the following categories:

- Frameworks in general from various standards and organizations like NIST, ISO, COBIT etc.
- Government strategies/frameworks
- Other organizations' strategies/frameworks/Smart City Maturity Models/ Frameworks

This information can be used as a live reference with further updates on existing and upcoming relevant frameworks and standards, as per need.

Frameworks

Sl. No.	Framework	Additional Info/Reference
1	NIST Special Publication 800-(Risk Assessment)	i. Assessing Security and Privacy Controls in Federal Information Systems and Organizations—Building Effective Assessment Plans—Revision 4: http://nvlpubs.nist.gov/nistpubs /SpecialPublications/NIST.SP.800-53Ar4 .pdf ii. Guide for Conducting Risk Assessments: http://csrc.nist.gov /publications/nistpubs/800-30-rev1 /sp800_30_r1.pdf iii. Guide for Applying the Risk Management Framework to Federal Information Systems : http://csrc.nist .gov/publications/nistpubs/800-37-rev1 /sp800-37-rev1-final.pdf iv. NIST Framework for Improving Critical Infrastructure Cybersecurity: http:// www.nist.gov/cyberframework/upload /cybersecurity-framework-021214-final .pdf
2	CMU's OCTAVE Allegro Method	http://www.cert.org/resilience/products -services/octave/octave-method.cfm
3	ISO 31000	https://www.iso.org/iso-31000-risk -management.html
4	ISACA's COBIT 5	www.isaca.org/cobit/
5	FAIR model- Factor Analysis of Information Risk	http://www.riskmanagementinsight.com /media/docs/FAIR_introduction.pdf
6	ISO 27005	https://www.iso.org/standard/56742.html

(Continued)

Sl. No.	Framework	Additional Info/Reference
7	CRAMM (CCTA Risk Analysis and Management Method)	i. A Qualitative Risk Analysis and Management Tool—CRAMM: http://www.sans.org/reading-room/whitepapers/auditing/qualitative-risk-analysis-management-tool-cramm-83 ii. Hospital Cybersecurity Risk Assessment Maturity Model (HCRAMM): https://dspace.library.uu.nl/handle/1874/298581
8	EBIOS (Expression of Needs and Identification of Security Objectives)—France	http://www.ssi.gouv.fr/en/the-anssi/publications-109/methods-to-achieve-iss/ebios-2010-expression-of-needs-and-identification-of-security-objectives.html
9	SOMAP	http://softlayer-ams.dl.sourceforge.net/project/somap/Infosec%20Risk%20Assessment%20Guide/Version%201.0/somap_guide_v1.0.0.pdf
10	ISO/IEC 30101:2009 Risk management—Risk assessment techniques	https://www.iso.org/standard/51073.html
11	Identity Ecosystem Framework	https://www.idecosystem.org/sites/default/files/IEF_Conceptual_Overview2.pdf
12	Federal Identity, Credential, and Access Management (FICAM) Roadmap and Implementation Guidance	http://www.idmanagement.gov/sites/default/files/documents/FICAM_Roadmap_and_Implementation_Guidance_v2%200_20111202_0.pdf
13	ISACA Adaptive Access framework	https://www.isaca.org/Groups/Professional-English/access-control/GroupDocuments/ISSA%2520Adaptive%2520Access%2520Framework%2520Whitepaper%2520Final%2520v1.docx
14	FRAP: Facilitated Risk Assessment Process	http://task.to/Portals/11/Resources/SpeedTalks08/2_rbeggs_frap.pdf
15	COSO ERM—Integrated Framework	https://www.coso.org/Pages/erm-integratedframework.aspx

(*Continued*)

Sl. No.	Framework	Additional Info/Reference
16	FRAMES—A Risk Assessment Framework for e-Services	http://www.ejeg.com/issue/download.html?idArticle=19
17	PCI DSS Version 3.2— PCI Security Standards Council	https://www.pcisecuritystandards.org/pdfs/PCI_DSS_Resource_Guide_(003).pdf
18	CORAS—A platform for risk analysis of security critical systems (European Commission R&D project)	http://telemed.no/coras-a-platform-for-risk-analysis-of-security-critical-systems.44416-247954.html
19	Federal Financial Institutions Examination Council (FFIEC) Authentication Guidance	i. https://chapters.theiia.org/western-new-york/ChapterDocuments/FFIEC%2520Authentication%2520Guidance.pptx ii. Supplement to Authentication in an Internet Banking Environment: http://www.ffiec.gov/pdf/auth-its-final%206-22-11%20(ffiec%20formated).pdf iii. FFIEC 's Cybersecurity Assessment: https://www.ffiec.gov/pdf/cybersecurity/2014_June_FFIEC-Cybersecurity-Assessment-Overview.pdf
20	The Orange Book: Management of Risk—Principles and Concepts	www.hm-treasury.gov.uk/d/orange_book.pdf
21	Cyber Risk Analysis Tool—MONARC— Luxembourg	http://visit.a-sit.at/praesentationen/petitgenet_Pr%C3%A9s_Viennes_CASES-v0.2.pdf
22	MEHARI— developed by a French association of information security professionals called CLUSIF	https://www.clusif.asso.fr/en/production/mehari/download.asp
23	IT-Grundschutz Germany	http://www.tenable.com/sc-dashboards/it-grundschutz-bsi-100-2-dashboard

(Continued)

Sl. No.	Framework	Additional Info/Reference
24	Information Security Assessment & Monitoring Method (ISAMM)—from European Union Agency for Network and Information Security	http://rm-inv.enisa.europa.eu/methods/m_isamm.html
25	STORM—RM	https://www.ictprotect.com/main/tool
26	OCEG GRC Capability Model (Red Book)	https://go.oceg.org/grc-capability-model-red-book
27	OCEG GRC ASSESSMENT Tools (Burgundy Book)	https://go.oceg.org/grc-assessment-tools-burgundy-book

Government Strategies/Frameworks

Sl. No.	Framework	Additional Info/Reference
1	National Strategy for Critical Infrastructure—Canada	https://www.publicsafety.gc.ca/cnt/rsrcs/pblctns/srtg-crtcl-nfrstrctr/index-eng.aspx
2	M-A-G-E-R-I-T Risk Assessment Methodology—Spanish government	http://www.seap.minhap.gob.es/dms/en/publicaciones/centro_de_publicaciones_de_la_sgt/Monografias0/parrafo/MAGERIT_3/MAGERIT_v_3_-book_1_method_PDF_NIPO_630-14-162-0/MAGERIT_v_3_%20book_1_method_PDF_NIPO_630-14-162-0.pdf
3	UK Information Management and Risk Assessment framework	http://www.nationalarchives.gov.uk/information-management/manage-information/polic y-process/digital-continuity/risk-assessment/
4	Information Assurance Maturity Model—UK	www.cesg.gov.uk/products_services/iacs/iamm/index.shtml

(Continued)

Sl. No.	Framework	Additional Info/Reference
5	Hawaii Government— Information Assurance and Cyber Security	https://oimt.hawaii.gov/wp-content /uploads/2012/09/Governance_Info -Assurance_Cyber-Security.pdf
6	NATIONAL STRATEGY FOR TRUSTED IDENTITIES IN CYBERSPACE	http://www.whitehouse.gov/sites /default/files/rss_viewer /NSTICstrategy_041511.pdf
7	Australian Government e-Authentication Framework (AGAF)—for business	http://www.finance.gov.au/agimo -archive/__data/assets/pdf _file/0013/42052/-7-Robyn_Fleming -AGAF.pdf
8	European Network and Information Security Agency (ENISA) Guidebook on National Cyber Security Strategies	https://www.enisa.europa.eu /publications/ncss-good -practice-guide
9	The UK Cyber Security Strategy—Protecting and promoting the UK in a digital world, Cabinet Office, United Kingdom	https://www.gov.uk/government /uploads/system/uploads/attachment _data/file/60961/uk-cyber-security -strategy-final.pdf
10	National e-Authentication Framework—Australia	http://www.finance.gov.au/files/2012/04 /NeAFFramework.pdf

Other Organizations' Strategies/Frameworks

Sl. No.	Framework/Strategies	Additional Info/Reference
1	Lloyds 360 Digital Risk Report	http://www.lloyds.com/~/media/lloyds /reports/360/360%20digital/lloyds_360 _digital_risk_report%20(2).pdf
2	Assessing & Managing Digital Risk-International Underwriting Association	http://trygstad.rice.iit.edu:8000/Articles /Assessing&ManagingDigitalRisk-Inte rnationalUnderwritingAssociation .pdf
3	Electronic Security: Risk Mitigation in Financial Transactions—World Bank	http://elibrary.worldbank.org/doi /pdf/10.1596/1813-9450-2870

(*Continued*)

Sl. No.	Framework/Strategies	Additional Info/Reference
4	Consumer-Perceived Risk in E-Commerce Transactions	http://www.som.buffalo.edu /isinterface/papers/Consumer -Perceived%20Risk%20in%20 E-Commerce.pdf
5	A Risk Assessment Framework for Mobile Payments	http://www.rogerclarke.com/EC /MP-RAF.html
6	Intel's Threat Agent Risk Assessment Framework	http://www.intel.in/content/dam/www /public/us/en/documents/solution -briefs/risk-assessments-maximize -security-budgets-brief.pdf
7	Threat Assessment & Remediation Analysis— MITRE Corporation	Threat Assessment & Remediation Analysis (TARA) is an engineering methodology to identify, prioritize, and respond to cyber threats through the application of countermeasures that reduce susceptibility to cyber attack. http://www.mitre.org/sites/default/files /pdf/11_4982.pdf
8	SANS	http://www.sans.org/reading-room /whitepapers/auditing/introduction -information-system-risk -management-1204
9	PWC—A practical guide to risk assessment	http://www.pwc.com/en_US/us/issues /enterprise-risk-management/assets /risk_assessment_guide.pdf
10	First Data e-commerce fraud risk assessment	http://www.firstdata.com/downloads /thought-leadership/ecommfraudwp .pdf
11	Risk Management for e-Business	http://revistaie.ase.ro/content/43/16 -nastase.pdf
12	Risk-Based E-Business Testing—Risks and Test Strategy	http://gerrardconsulting.com/papers /articles/EBTestingPart1.pdf
13	Microsoft Compliance Framework for Online Services—Global	http://cdn.globalfoundationservices .com/documents/Microsoft ComplianceFramework1009.pdf

Smart City Maturity Models/Frameworks

Sl. No.	Cities/Models/Framework /Strategies	Additional Info/Reference
1	Smart City Maturity Model—Scottish Cities Alliance	https://www.scottishcities.org.uk/site /assets/files/1103/smart_cities _readiness_assessment_-_guidance _note.pdf
2	Smart Cities Readiness Guide—Smart Cities Council	https://smartcitiescouncil.com /resources/smart-cities-readiness -guide
3	PAS 181—Smart Cities Framework (BSI)	https://shop.bsigroup.com/upload /Smart_cities/BSI-PAS-181-executive -summary-UK-EN.pdf
4	Barcelona Smart City	http://meet.barcelona.cat/en/
5	Copenhagen Smart City	www.copcap.com/set-up-a-business /key-sectors/smart-city
6	Smart Nation Singapore	www.smartnation.sg/
7	Amsterdam Smart City	https://amsterdamsmartcity.com/
8	Chicago Smart City	www.smartchicagocollaborative.org/
9	Glasgow Future City	futurecity.glasgow.gov.uk/
10	New York Smart Equitable City	https://www1.nyc.gov/site/forward /innovations/smartnyc.page
11	India Smart Cities Mission	http://smartcities.gov.in/content/
12	Toronto Smart City	https://www.toronto.ca/community -people/get-involved/smart-cities -challenge/
13	Vienna Smart City	https://smartcity.wien.gv.at/site/en/
14	Sydney Smart City	http://sydneymatters.org.au /wp-content/uploads/2016/08/Sydney -Matters-Smart-City-Policy.pdf
15	Berlin Smart city	https://www.berlin-partner.de/en /the-berlin-location/smart-city-berlin/
16	Songdo Smart City	https://newsroom.cisco.com/songdo

Appendix 3: Various Global Initiatives Related to the IoT and Smart Cities

Worldwide businesses, technical and policy-research institutes, standards-creating bodies and think tanks are working on realizing the potential of IoT technology and smart cities. Some of these organizations are highlighted below, which are leading these initiatives in various aspects like security, privacy, standardization of technology, societal impacts and so on.

Cloud Security Alliance

Cloud Security Alliance (CSA) is the world's leading organization dedicated to defining and raising awareness of best practices to help ensure a secure cloud-computing environment.

The CSA IoT Working Group focuses on understanding the relevant use-cases for IoT deployments and defining actionable guidance for security practitioners to secure their implementations.

The working group is chartered to research the following areas:

- Analysis of IoT implementation use-cases in various industries
- Best practices for securing IoT implementations
- Mapping of IoT security controls to the cloud controls matrix
- Identifying threats to IoT devices and implementations
- Identifying gaps in standards coverage for IoT security
- Identifying gaps in technology solutions for IoT security
- Research into new methods for securing the IoT
- Coordination with other CSA working groups and with external security organizations to de-conflict and jointly define cyber security controls for the IoT

Figure A3.1   Cloud Security Alliance.

Cloud Security Alliance. (n.d.). Retrieved from https://cloudsecurityalliance.org/group/internet-of-things/#_downloads

- Securing cloud infrastructure and services that support the IoT
- Securing edge devices to remove the threat of follow-on compromises to the enterprise
- Solutions for auditing, identity and access management, authentication, inventory management, privacy and risk management of the IoT.

Smart Cities for All

G3ict and World Enabled launched the Smart Cities for All initiative to define the state of ICT accessibility in smart cities worldwide. Its focus is to eliminate the digital divide for persons with disabilities and older persons in smart cities around the world. They are partnering with leading organizations and companies to create and deploy the tools and strategies needed to build more inclusive smart cities.

They have developed the "Smart Cities for All Toolkit" that contains four tools to help smart cities worldwide include a focus on ICT accessibility and the digital inclusion of persons with disabilities and older persons.

ETSI

The European Telecommunications Standards Institute (ETSI) is an independent, not-for-profit and recognized regional standards body in Europe that deals with

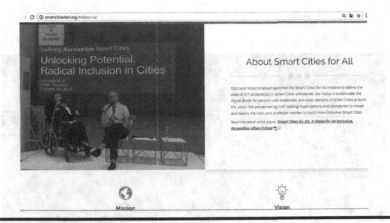

Figure A3.2   Smart Cities for All.

Smart Cities for All. (n.d.). Retrieved from http://smartcities4all.azurewebsites.net
/#about-us

telecommunications, broadcasting and other electronic communications networks
and services.

ETSI is involved in standardizing a wide range of technologies that work
together to connect things in the IoT and smart cities. ETSI is helping find the
necessary radio spectrum for connecting things wirelessly in the IoT. They have
also developed a baseline specification using 'surface mount technology' that will
simplify the integration of modules from different manufacturers in a wide range
of M2M applications.

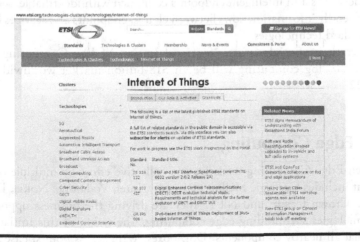

Figure A3.3   ETSI.

ETSI - Internet of Things. (n.d.). Retrieved from https://www.etsi.org/technologies-
clusters/technologies/internet-of-things

Figure A3.4 OpenFog Consortium.
OpenFog Consortium. (n.d.). Retrieved from https://www.openfogconsortium.org/

OpenFog Consortium

The OpenFog Consortium aims to drive industry and academic leadership in fog computing & networking architecture, testbed development, and interoperability and composability deliverables that seamlessly bridge the cloud-to-things continuum.

The consortium's work is centered around creating a framework for efficient and reliable networks and intelligent endpoints combined with identifiable, secure and privacy-friendly information flows between clouds, endpoints and services based on open-standard technologies.

As per their special report, "Size and Impact of Fog Computing Market (2017-2022)," fog computing will be an $18-billion market worldwide by the year 2022.

Open Edge Computing

The Open Edge Computing initiative is a collective effort by multiple organizations that includes Carnegie Mellon University, Intel, Vodafone, Nokia, Deutsche Telekom, Crown Castle, and NTT. They are driving the business opportunities and technologies surrounding edge computing. The vision of this initiative is to ensure that all nearby components (DSL-boxes, WiFi access points, base stations) offer resources through open and standardized mechanisms to any application, device, or sensor to enable computation at the edge.

They are building 'Living Edge Lab,' an open and flexible lab for edge computing.

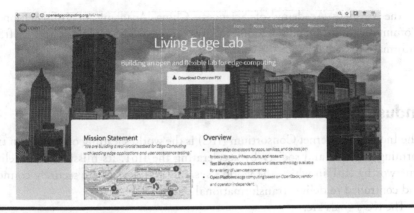

Figure A3.5   Open Edge Computing.

Open Edge Computing. (n.d.). Retrieved from http://openedgecomputing.org/lel.html

EDGE Computing Consortium

The EDGE Computing Consortium (ECC) serves as an edge-computing industry cooperative platform, which promotes open cooperation in the Operational Technology (OT) and ICT fields, nurtures the industry's best application practices and advances the sound and sustainable development of the edge-computing industry.

To promote in-depth industry coordination, accelerate innovation and boost the application of edge computing, six industry entities have joined together to establish the ECC: Huawei Technologies, Shenyang Institute of Automation

Figure A3.6   EDGE Computing Consortium.

EDGE Computing Consortium. (n.d.). Retrieved from http://en.ecconsortium.org/index/index,htlm/#page4

of the Chinese Academy of Sciences, China Academy of Information and Communications Technology, Intel Corporation, ARM Holdings and iSoftStone Information Technology.

Industrial Internet Consortium

The Industrial Internet Consortium (IIC) is the world's leading organization transforming business and society by accelerating the IIoT. Its mission is to deliver a trustworthy IIoT in which the world's systems and devices are securely connected and controlled to deliver transformational outcomes.

The IIC's goals are:

- To drive innovation through the creation of new industry use-cases and test-beds for real-world applications
- To define and develop the reference architecture and frameworks necessary for interoperability
- To Influence the global development standards process for Internet and industrial systems
- To facilitate open forums to share and exchange real-world ideas, practices, lessons and insights
- To build confidence around new and innovative approaches to security.

Figure A3.7   Industrial Internet Consortium.

Industrial Internet Consortium. (n.d.). Retrieved from https://www.iiconsortium .org/index.htm

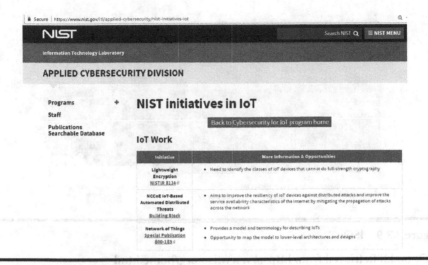

Figure A3.8 NIST.

NIST initiatives in IoT. (n.d.). Retrieved from https://www.nist.gov/itl/applied
-cybersecurity/nist-initiatives-iot

NIST

The National Institute of Standards and Technology (NIST) was founded in 1901
and is now part of the U.S. Department of Commerce.

NIST's Cybersecurity for the Internet of Things program supports the development and application of standards, guidelines and related tools to improve
the cybersecurity of connected devices and the environments in which they are
deployed. By collaborating with stakeholders across governments, industries, international bodies and academia, the program aims to cultivate trust and promote
U.S. leadership in the IoT.

ISO

International Organization for Standardization (ISO) is an independent, non-
governmental international organization with a membership of 162 national standards bodies.

Through its members, it brings together experts to share knowledge and develop
voluntary, consensus-based, market-relevant international standards that support
innovation and provide solutions to global challenges.

ISO/IEC JTC 1/SC 41 works specifically on the standardization of the IoT and
related technologies. A key initiative of ISO/IEC JTC 1/SC 27 is the standardization effort in the security and privacy aspects of the IoT and smart cities. ISO/IEC

Figure A3.9   ISO.

ISO. (n.d.). Retrieved from https://www.iso.org/home.html

JTC1 WG11 serves as the focus of and proponent for JTC 1's smart cities standardization program.

Zigbee Alliance

Zigbee Alliance is an open, non-profit association of members that creates IoT standards for wireless connectivity. Since 2002, the Zigbee Alliance and its member companies have created standards, certification programs and tools to develop interoperable products for the low-power wireless IoT.

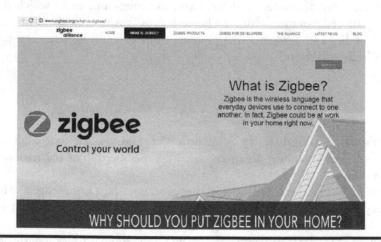

Figure A3.10   Zigbee Alliance.

Zigbee Alliance. (n.d.). Retrieved from http://www.zigbee.org/what-is-zigbee/

The Zigbee Alliance offers a Software Test Tool as part of the Zigbee Certification Program to assist with Zigbee device development, improve the testing and verification processes and to reduce integral cost and time to market.

oneM2M

oneM2M is the global standards initiative for machine-to-machine communications and the IoT.

The purpose and goal of oneM2M is to develop technical specifications which address the need for a common M2M service layer that can be readily embedded within various hardware and software and relied upon to connect the myriad of devices in the field with M2M application servers worldwide.

A critical objective of oneM2M is to attract and actively involve organizations from M2M-related business domains such as telematics and intelligent transportation, healthcare, utilities, industrial automation and smart homes.

IEEE

IEEE is the world's largest technical professional organization dedicated to advancing technology for the benefit of humanity.

The IEEE Internet of Things Initiative is one of the IEEE's most important, multi-disciplinary, cross-platform initiatives. The mission of the IEEE IoT Initiative is to serve as the gathering place for the global technical community working on

Figure A3.11   oneM2M.

Standards for M2M and the Internet of Things. (n.d.). Retrieved from http://www .onem2m.org/technical/published-drafts

Figure A3.12   IEEE.

IEEE Internet of Things: Home. (n.d.). Retrieved from https://iot.ieee.org/

the Internet of Things; to provide the platform where professionals learn, share knowledge, and collaborate on this sweeping convergence of technologies, markets, applications, and the Internet, and together change the world.

ITU-T

ITU is the United Nations' specialized agency for ICTs.

ITU's Telecommunication Standardization Sector (ITU-T) assembles experts from around the world in study groups to develop international standards known as 'ITU-T recommendations' which act as defining elements in the global infrastructure of ICTs.

ITU-T's Study Group 20 (SG20) is working to address the standardization requirements of IoT technologies, with an initial focus on IoT applications in smart cities and communities.

SG20 develops international standards to enable the coordinated development of IoT technologies, including machine-to-machine communications and ubiquitous sensor networks.

ISOC

The Internet Society (ISOC) is a global cause-driven organization governed by a diverse board of trustees that is dedicated to ensuring that the Internet stays open, transparent and defined by you.

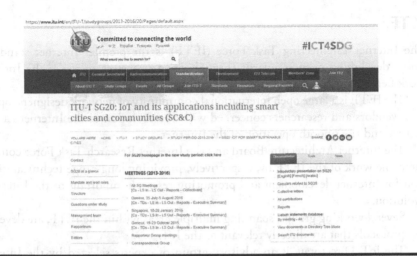

Figure A3.13 ITU-T.

ITU-T SG20: IoT and its applications including smart cities and communities (SC&C). (n.d.). Retrieved from https://www.itu.int/en/ITU-T/studygroups/2013-2016/20 /Pages/default.aspx

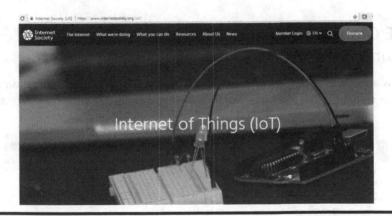

Figure A3.14 ISOC.

Internet of Things (IoT) | Internet Society. (n.d.). Retrieved from https://www .internetsociety.org/iot/

To understand the opportunities and challenges associated with the IoT, ISOC published "The Internet of Things: An Overview—Understanding the Issues and Challenges of a More Connected World," a whitepaper that examines many important aspects of the IoT.

In April 2017, the Online Trust Alliance (OTA) became an initiative of ISOC. Among the OTA's activities is their IoT Trust Framework.

IETF

The Internet Engineering Task Force (IETF) is the premier Internet standards body, developing open standards through open processes to make the Internet work better.

The IETF is a large open international community of network designers, operators, vendors and researchers concerned with the evolution of the Internet's architecture and the smooth operation of the Internet.

The Internet Architecture Board and the Internet Research Task Force complement the work of the IETF by, respectively, providing long-range technical direction for Internet development and promoting research important to the Internet's evolution.

Several working groups, spanning multiple areas within the IETF, are developing protocols that are directly relevant to the IoT.

The IoT Directorate is an advisory group of experts selected by the Internet Area directors and the Directorate Chairs. The main purpose of the Directorate is coordination within IETF IoT groups and increasing IETF IoT standards visibility to external Standard Developing Organizations (SDOs), alliances, and other organizations.

IEEE Global Initiative on Ethics of Autonomous and Intelligent Systems

This is an initiative of the IEEE Standards Association. It is an incubation space for new standards and solutions, certifications and codes of conduct and consensus

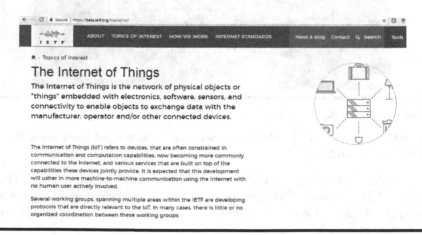

Figure A3.15 IETF.

IETF | Internet of things. (n.d.). Retrieved from https://www.ietf.org/topics/iot/

Figure A3.16   IEEE Global Initiative on Ethics of Autonomous and Intelligent Systems.

IEEE Global Initiative on Ethics of Autonomous and Intelligent Systems. (n.d.). Retrieved from https://standards.ieee.org/develop/indconn/ec/autonomous _systems.html

building for ethical implementation of intelligent technologies, so that these technologies are advanced for the benefit of humanity.

They have developed the principles for Ethically Aligned Design: A Vision for Prioritizing Human Well-being with Autonomous and Intelligent Systems (A/IS). The goal of Ethically Aligned Design is to advance a public discussion about how we can establish ethical and social implementations for intelligent and autonomous systems and technologies, aligning them to defined values and ethical principles that prioritize human well-being in a given cultural context.

ID3

The Institute for Data Driven Design (ID3) is a research and educational nonprofit, headquartered in Boston, Massachusetts. Its mission is to develop a new social ecosystem of trusted, self-healing digital institutions. This endeavor seeks to address the severe structural limitations of existing institutions by empowering individuals to assert greater control over their data, online identities and authentication, and in so doing, enable them to design and deploy a new generation of trusted digital institutions and services globally.

Open Mustard Seed (OMS) is an open-source framework from ID3 for developing and deploying secure and trusted cloud-based and mobile applications. OMS integrates a stack of technologies including hardware-based trusted execution environments, blockchain 2.0, machine learning and secure mobile and cloud based computing.

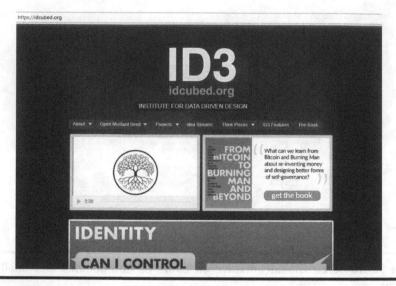

Figure A3.17 ID3.

Institute for Data-Driven Design. (n.d.). Retrieved from http://www.idcubed.org/

OGC

The Open Geospatial Consortium (OGC) is an international not-for-profit organization of over 522 companies, government agencies and universities committed to making quality open standards for the global geospatial community. These standards are made through a consensus process and are freely available for anyone to use to improve the sharing of the world's geospatial data.

The OGC Interoperability Program initiative on the IoT aims to help develop the consensus-standards infrastructure necessary to achieve the full societal, economic and scientific benefits of location information in mobile applications worldwide.

OGC has a SmartCities Domain Working Group (DWG). The basic elements of DWG that fit in the scope of the OGC model and domain are applications and services based on both mobile and static networks to support the data, application and monitoring requirements of smart cities, its applications, its datasets and its history and processes.

G3ict

The Global Initiative for Inclusive Information and Communication Technologies (G3ict) is an advocacy initiative launched in December 2006 by the United Nations Global Alliance for ICT and Development, in cooperation with the Secretariat for the Convention on the Rights of Persons with Disabilities at the United Nations

Figure A3.18 OGC.

GeoMobile and Internet of Things | OGC - Open Geospatial Consortium. (n.d.). Retrieved from www.opengeospatial.org/projects/initiatives/geomobilecd

Department of Economic and Social Affairs (DESA). Its mission is to facilitate and support the implementation of the dispositions of the Convention on the Rights of Persons with Disabilities on the accessibility of ICTs and assistive technologies. G3ict relies on an international network of ICT accessibility experts to develop and

Figure A3.19 3Gict.

G3ict Smart Cities Initiative. (n.d.). Retrieved from http://www.g3ict.org/resource _center/g3ict_smart_cities_initiative

promote good practices, technical resources and benchmarks for ICT accessibility advocates around the world.

Smart Cities for All is a global strategy for digital Inclusion proposed by G3ict and World Enabled.

IGF

The Internet Governance Forum (IGF) serves to bring people together from various stakeholder groups as equals in discussions on public policy issues relating to the Internet. While there is no negotiated outcome, the IGF informs and inspires those with policy-making power in both the public and private sectors. At their annual meeting, delegates discuss, exchange information and share good practices with each other. The IGF facilitates a common understanding of how to maximize Internet opportunities and address risks and challenges that arise.

The Secretary-General of the United Nations established the Advisory Group (now referred to as the Multistakeholder Advisory Group, or MAG). Its purpose is to advise the Secretary-General on the program and schedule of IGF meetings. The MAG is comprised of 55 members from governments, the private sector and civil society, including representatives from the academic and technical communities.

Figure A3.20   IGF.

Internet Governance Forum. (n.d.). Retrieved from https://www.intgovforum.org/multilingual/

Figure A3.21   Data & Society.
About | Data & Society. (n.d.). Retrieved from https://datasociety.net/about/

Data & Society

Data & Society is a research institute in New York City that is focused on the social and cultural issues arising from data-centric and automated technologies.

To provide frameworks that can help society address emergent tensions, Data & Society is committed to identifying thorny issues at the intersection of technology and society, providing and encouraging research that can ground informed, evidence-based public debates, and building a network of researchers and practitioners who can anticipate issues and offer insight and direction.

World Economic Forum

The World Economic Forum, committed to improving the state of the world, is the international organization for public-private cooperation.

The Forum engages the foremost political, business and other leaders of society to shape global, regional and industry agendas.

It was established in 1971 as a not-for-profit foundation and is headquartered in Geneva, Switzerland. It is independent, impartial and not tied to any special interests. The Forum strives in all its efforts to demonstrate entrepreneurship in the global public interest while upholding the highest standards of governance. Moral and intellectual integrity is at the heart of everything it does.

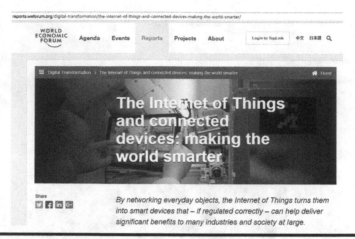

Figure A3.22   World Economic Forum.

The Internet of Things and connected devices - World Economic Forum. (n.d.). Retrieved from http://reports.weforum.org/digital-transformation/the-internet-of -things-and-connected-devices-making-the-world-smarter/

Smart Cities Council

The Smart Cities Council is an industry coalition formed to accelerate the move to smart, sustainable cities. It envisions a world where digital technology and intelligent design have been harnessed to create smart, sustainable cities with high-quality living and high-quality jobs. To tap into the transformative power of smart

Figure A3.23   Smart Cities Council.

Smart Cities Council | Teaming to build the cities of the future. (n.d.). Retrieved from https://smartcitiescouncil.com/

technologies, cities need a trusted, neutral advisor. The Smart Cities Council provides that help.

The Council helps cities become smarter through a combination of advocacy and action.

FPF

Future of Privacy Forum (FPF) is a non-profit organization that serves as a catalyst for privacy leadership and scholarship, advancing principled data practices in support of emerging technologies.

Figure A3.24   FPF.

Future of Privacy Forum. (n.d.). Retrieved from https://fpf.org/

Figure A3. 25   AIOTI.

AIOTI - SPACE | The Alliance for the Internet of Things Innovation. (n.d.). Retrieved from https://aioti.eu/

FPF brings together industry, academics, consumer advocates, and other thought leaders to explore the challenges posed by technological innovation and develop privacy protections, ethical norms and workable business practices.

FPF helps fill the void in the 'space not occupied by law' which exists due to the speed of technology development.

As 'data optimists,' FPF believes that the power of data for good is a net benefit to society, and that it can be well-managed to control risks and offer the best protections and empowerment to consumers and individuals.

AIOTI

The Alliance for Internet of Things Innovation (AIOTI) was initiated by the European Commission in 2015, with the aim to strengthen the dialogue and interaction among IoT players in Europe and to contribute to the creation of a dynamic European IoT ecosystem to speed up the take-up of the IoT.

Other objectives of the AIOTI include fostering experimentation, replication and deployment of IoT and supporting convergence and interoperability of IoT standards; gathering evidence on market obstacles for IoT deployment; and mapping and bridging global, EU and member states' IoT innovation activities.

Index

Page numbers followed by f and t indicate figures and tables, respectively.

Printed in the United States
by Baker & Taylor Publisher Services